Artificial Intelligence-based Cybersecurity for Connected and Automated Vehicles

Jordi Guijarro Olivares, Peter Hofmann, Petros Kapsalas, Jordi Casademont, Saber Mhiri, Nikos Piperigkos, Rodrigo Diaz, Bruno Cordero, Jordi Marias, Adrián Pino, Theocharis Saoulidis, Josep Escrig, Choi You Jun and Taesang Choi

now
the essence of knowledge

Published, sold and distributed by:
now Publishers Inc.
PO Box 1024
Hanover, MA 02339
United States
Tel. +1-781-985-4510
www.nowpublishers.com
sales@nowpublishers.com

Outside North America:
now Publishers Inc.
PO Box 179
2600 AD Delft
The Netherlands
Tel. +31-6-51115274

ISBN: 978-1-63828-060-6
E-ISBN: 978-1-63828-061-3
DOI: 10.1561/9781638280613

Suggested citation: Jordi Guijarro Olivares, Peter Hofmann, Petros Kapsalas, Jordi Casademont, Saber Mhiri, Nikos Piperigkos, Rodrigo Diaz, Bruno Cordero, Jordi Marias, Adrián Pino, Theocharis Saoulidis, Josep Escrig, Choi You Jun and Taesang Choi. (2022). *Artificial Intelligence-based Cybersecurity for Connected and Automated Vehicles.* Boston–Delft: Now Publishers

This book has received funding from the European Union's Horizon 2020 research and innovation programme under grant agreement No. 833611.

Table of Contents

EC Final Review Evaluation

"Project has delivered exceptional results with significant immediate or potential impact The project has achieved all of its objectives and milestones for the period and went much beyond the dissemination objectives.

The main goal of the project was the development of a more secure driving experience for the connected and automated vehicles and was built around four innovation pillars: the autonomous vehicle, the connected vehicle, the plug-in electrical vehicle and the remote control vehicle.

Significant realizations have been performed with respect to standardization activities, dissemination in academic, commercial and industrial fora and by creating synergies and liaisons with other running EU funded projects and initiatives. A final and successful demonstration found place on June 20th–21 in Langen (Germany), whose results have been clearly reported in the foreseen deliverable and presented during the panel review.

There was a very fruitful interaction with the Advisory board, who consisted of major relevant people in the field. The overall topic of CARAMEL is very relevant and important for the European competitiveness.

This work is highly important as is shown in the project, future cars will be digital, autonomous, and networked and hence vulnerable to digital attacks. It is important to continue the technical work to bring these steps to secure cars into the market, but also as important to take a look at the broader picture of the attack vectors that will be possible in the future world. The digital security of the whole system will be dependent on the organizational and technical security of the party providing the PKI for the secure platform. The role of organizational security should be studied further. At the same time, there is still more work that can be done to conceptually simplify the security threats that networked cars face – in order to help that understanding better penetrate the car design and manufacturing systems that are mostly not security oriented."

Overview

From padded dashboards to seat belts and from rear-view cameras to active safety measures. Nowadays, cars are becoming safer, smarter, and "greener" through connectivity, Artificial Intelligence (AI), and Machine Learning (ML), while cybersecurity aims for more sustainable safer roads with zero fatality. The CARAMEL project developed cybersecurity solutions for the new generation of cars: (i) autonomous cars, (ii) 5G connected vehicles (iii) electromobility and (iv) Remote control vehicles. CARAMEL applied a proactive method based on AI and ML techniques to mitigate cybersecurity-originated safety risks on roads. Considering the entire supply chain, CARAMEL developed innovative anti-hacking intrusion detection/prevention systems for the European automotive industry. So far, CARAMEL analyzed the security and privacy requirements of future mobility, identified the attack surface and modeled the potential threats, designed the overall architecture and defined the system specifications, elaborated use cases, and identified cyber threat detection and response techniques for all four project's pillars. Moreover, CARAMEL worked on enabling technologies and ML/AI techniques to detect and mitigate cyber threats to future mobility, in particular, innovative distributed PKI infrastructure, AI-based context-rich and context-aware solutions, holistic ML-based solutions based on the fusion of multiple data sources, and the CARAMEL backend to help mobility actors like road owners, traffic managers, etc. In addition, CARAMEL's design of an innovative anti-hacking device is a key solution to protect future roads and vehicles.

The damaging effects of cyberattacks on Cooperative Connected and Automated Mobility (CCAM) can be tremendous. Such as the damage to the reputation of vehicle manufacturers, the increased denial of customers to adopt CCAM, the loss of working hours, material damages, increased environmental pollution due e.g., to traffic jams or malicious modifications in sensors' firmware, and the danger for human lives. CARAMEL proactively addresses modern vehicle cybersecurity challenges by using advanced AI and ML techniques while seeking methods to mitigate associated safety risks. Addressing cybersecurity considerations for current CCAM vehicles, well-established ICT sector methodologies will be adopted, allowing vulnerabilities and potential cyberattack impacts assessment. Past automotive industry initiatives and cybersecurity projects have reached security assurance frameworks for networked vehicles, but several newly introduced technological dimensions like 5G, autopilots, and smart charging of Electric Vehicles (EVs) introduce cybersecurity gaps, not been fully addressed yet.

The project has submitted all its deliverables and achieved all of its challenges, objectives, and milestones. The progress and the partner's commitment ensured the project's successful use cases implementation, despite the added complexity of deploying scenarios in real environments.

The CARAMEL project strives to apply cybersecurity methodologies to the detection and mitigation of cybersecurity threats in the automotive domain. Innovation action in the project focuses on four selected pillars: In pillar 1 (Autonomous Mobility) the project developed innovative technologies to detect attacks against the sensors in the vehicle, e.g. attacks using generative adversarial networks to disturb the object detection algorithms in the car. To this end, a novel intrusion detection system – the anti-hacking device – was integrated directly into the vehicle. The anti-hacking device uses machine-learning technology to detect attacks. Due to its unique design, it will be easy and safe to update these detection algorithms regularly to counter novel attacks in a short time. This research made autonomous or semi-autonomous driving more secure against advanced attack scenarios and will drive the acceptance and adoption of these innovations by the general public in Europe.

In pillar 2 (Connected Mobility) the project developed advanced attack detection technologies for the connected vehicle, especially the V2X protocols between the car and other cars and roadside units. Additionally, partners implemented innovative threat detection mechanisms for direct integration into onboard units (controllers built into the car). Furthermore, the project developed an innovative bridge technology to make competing 802.11p and C-V2X networking technologies compatible using MEC devices or roadside units. In pillar 3 (Electromobility) partners implemented security mechanisms to protect against wide-scale attacks targeting the European EV charging infrastructure: Charging stations are part of the European critical infrastructure, and attacks that cripple this infrastructure could adversely affect the flow of passengers and goods in Europe, the project implemented machine learning models to detect attacks on the level of the whole eCharging backend infrastructure so that early mitigations can be implemented.

In pillar 4 (Remote Control Vehicle) the Korean partners, who still has one year of project based on Korean funding, continued working on the implementation of the remote control vehicle based on the mmWAVE (23GHz) use case, building a data processing architecture and developing a Malicious Traffic Detection Solution LSTM (Long-Short Term Memory)-based cyber-attack anomaly prediction/detection.

Under the framework of the CARAMEL project, researchers have developed several simulation demonstrations to present the results of their research and integrations in mitigating cyber-attacks on autonomous vehicles. Over 14 scientific papers and articles have been published in international conferences and high-impact journals. Additionally, the consortium ensured a strong presence at trade fairs and exhibitions, including the IoT Solutions World Congress 2022 celebrated in Barcelona, and released a series of video demonstrations. In June 2022, the project achieved a significant milestone at Panasonic's facilities in Langen (Hessen), Germany, so the project results were successfully presented to a panel of external observers. THE H2020 CARAMEL project comes to an end after a journey of 33 months in which all the partners involved worked together towards the same main objective: the development of a more secure driving experience for connected and automated vehicles.

The COVID-19 pandemic undoubtedly impacted CARAMEL's activities, but we managed to meet the challenge thanks to digital collaboration and an enthusiastic, consistent, and constant team. As it was remarked by the European Commission representatives and resulted in receiving extra congratulations.

All these CARAMEL activities made the evolution of Automotive technology to a more connected and software-driven future more secure, therefore driving public acceptance of these technologies in Europe as well as giving European companies a competitive advantage.

Acknowledgements

This project has received funding from the European Union's Horizon 2020 research and innovation programme under grant agreement No. 833611.

Thanks **Pouria Khodashenas** for your vision and your effort in making it possible.

EC Project Officer: Hubert SCHIER

EC Project Reviewers:

Mikko Johannes SÄRELÄ – TEKNIIKAN AKATEEMISET RY
An BRAEKEN – Vrije Universiteit Brussel

External Advisory Board:

Jesús Alonso-Zarate – 5GPPP
Pedro Dias Rodrigues – EDP
Antonio Lopez Pena – CVC
Natalie Bertels – KULEUVEN
Brigitte Lonc – Renault
Johanna Tzanidaki – Ertico

Security Advisory Board:

Charalampos Sergiou – UCY
Jose Maria Blanco Navarro – PROSEGUR
Michal Choras – ITTI

Final demo observers

Tamara Djukic	ERTICO	Netherlands
Pedro Rodrigues	EDG	Portugal
YoungJun MOON	KOTI	KR
JongMyung KIM	KTL	KR

Other related projects

PID2019-106808RA-I00 and PID2020-112675RB-C43 funded by MCIN/AEI/ 10.13039/501100011033.

INTEGRA CER-20211031 funded by CDTI Cervera.

List of Acronyms

- EU: European Union
- PKI: Public key infrastructure
- AI: Artificial Intelligence
- ML: Machine Learning
- 5G: the fifth-generation technology standard for broadband cellular networks
- CCAM: Cooperative, connected and automated mobility
- ICT: information and communications technology
- EV: Electric Vehicles
- V2X: Vehicle-to-Everything
- C-V2X: Cellular vehicle-to-everything
- MEC: Multi-access Edge Computing
- LSTM: Long-Short Term Memory
- IoT: Internet of Things
- SAB: Security Advisory Board
- LTE: Long Term Evolution
- HSM: hardware secure module
- TCOS: Telekom Card Operating System
- SOC: security operation center
- GPS: The Global Positioning System
- ADAS: Advanced driver-assistance systems
- IP: Internet Protocol
- LiDAR: Light Detection and Ranging
- HMI: Human Machine Interface
- PPS: Path planning system
- VANET: vehicular ad hoc network
- RADAR: radio detection and ranging

- CL: Cooperative Localization
- MLE: Maximum Likelihood Estimation
- CVX: Matlab-based modeling system for convex optimization
- RCGCL: Robust Graph based Centralized Cooperative Localization
- KM: kinematic model
- LMSE: Localization Mean Square Error
- ROC: Receiver Operating Characteristics
- AUC: Area Under Curve
- KF: Kalman filtering
- WT: wavelet transform
- EMD: empirical mode decomposition
- NCSR: non-locally centralized sparse representation
- MRF: Markov random field
- WNNM: weighted nuclear norm minimization
- LSSC: learned simultaneous sparse coding
- CSF: cascade of shrinkage fields
- TNRD: trainable nonlinear reaction diffusion
- GHEP: gradient histogram estimation and preservation
- SSIM: Structural similarity index measure
- FGSM: fast gradient symbol method
- PGD: Projected Gradient Descent
- IoU: Intersection over Union
- GNSS: Global navigation satellite system
- ITS: Intelligent Transportation Systems
- ETSI: European Telecommunications Standards Institute
- WAVE: ITS-G5 suite in Europe and the Wireless Access in Vehicular Environments
- OBU: On Board Unit
- MEC: Multi-Access Edge Computing
- UAV: Unmanned Aerial Vehicles
- V2V: Vehicle-to-Vehicle
- V2I2V: Vehicle-to-Infrastructure-to-Vehicle
- AT: Authorisation Tickets
- BTP: Basic Transport Protocol
- GN: GeoNetworking protocol
- DENM: Decentralized Event Notification Messages
- SPATEM: Signal Phase And Timing Extended Message
- MAPEM: MAP Extended Message
- CRL: certificate revocation lists
- NIC: Network Interface Card

- IVN: In-Vehicle Network
- RSDB: Real Simultaneous Dual Band
- CAN: Controller Area Network
- RSU: Road Side Units
- OS: Operating System
- DFS: Dynamic Frequency Selection
- eNBs: evolved Node B
- EPC: Evolved Packet Core
- PCRF: Policy and Charging Rules Function
- APN: Access Point Name
- ROI: Region of Interest
- LDM: Local Dynamic Map
- MQTT: Message Queuing Telemetry Transport
- JSON: JavaScript Object Notation
- PoC: Proof of Concept
- IDS: Intrusion Detection Systems
- RCA: Root Certification Authority
- EC: Enrolment Certificates
- BC: Bootstrap Certificate
- BSP: Board Support Package
- ROM: Read Only Memory
- vEPC: virtual Evolved Packet Core
- EVSE: electric vehicle supply equipment
- PEVs: Plug-in Electrical Vehicles
- EVCSMS: EVCS management systems
- DSO: Distribution System Operator
- TSO: Transmission System Operators
- DDoS: Distributed Denial of Service
- IDPS: Intrusion Detection and Prevention System
- SIEM: Security information and event management
- APN: Access Point Name
- OCPP: Open Charge Point Protocol
- MV: MeterValues
- eMSP: e-Mobility Service Providers
- RCV: Release, Control and Validate
- eMBB: Enhanced Mobile Broadband
- URLLC: Ultra-Reliable Low Latency Communication
- BS: base station
- LVDS: Low Voltage Differential Signaling
- mmWAVE: Millimeter wave

- UE: User Equipment
- RSRP: Reference Signal Received Power
- IDM: Intelligent driving module
- ECM: engine control module
- BCM: Body control module
- ECU: electric control unit
- RNN: recurrent neural network

Introduction

Car safety has come a long way. From the first padded dashboard to seat belts and from rear-view cameras to active safety measures, technological advances are picking up speed. Nowadays, cars are becoming smarter and "greener" through connectivity and artificial intelligence, and cybersecurity is emerging as a new concern able to stop such huge potential for more sustainable safer roads with zero fatality. The CARAMEL project developed cybersecurity solutions for the new generation of cars: (i) autonomous cars, (ii) 5G connected vehicles (iii) electromobility and (iv) Remote control vehicles (Korean Partners) (see Figure 1).

The project applied a proactive method based on artificial intelligence and machine learning techniques to mitigate cybersecurity-originated safety risks on roads. Considering the entire supply chain, CARAMEL developed innovative anti-hacking intrusion detection/prevention systems for the European automotive industry. So far, the project analyzed the security and privacy requirements of the future mobility, identified the attack surface and modeled the potential threats, designed the overall architecture and defined the system specifications, elaborated use cases and identified cyberthreats detection and response techniques for all four pillars of the project. Moreover, the project also worked on the enabling technologies and ML/AI techniques to detect and mitigate cyberthreats on the future mobility, in particular, innovative distributed PKI infrastructure, AI-based context-rich and context-aware solution, holistic ML based solution based on the fusion of multiple data sources and the CARAMEL backend to help mobility actors like road owners, traffic managers, etc. In addition, the project designed an innovative anti-hacking device which is a key solution to protect future roads and vehicles.

The damaging effects of cyberattacks to an industry like the Cooperative Connected and Automated Mobility (CCAM) can be tremendous. One can mention for example the damage in the reputation of vehicle manufacturers, the

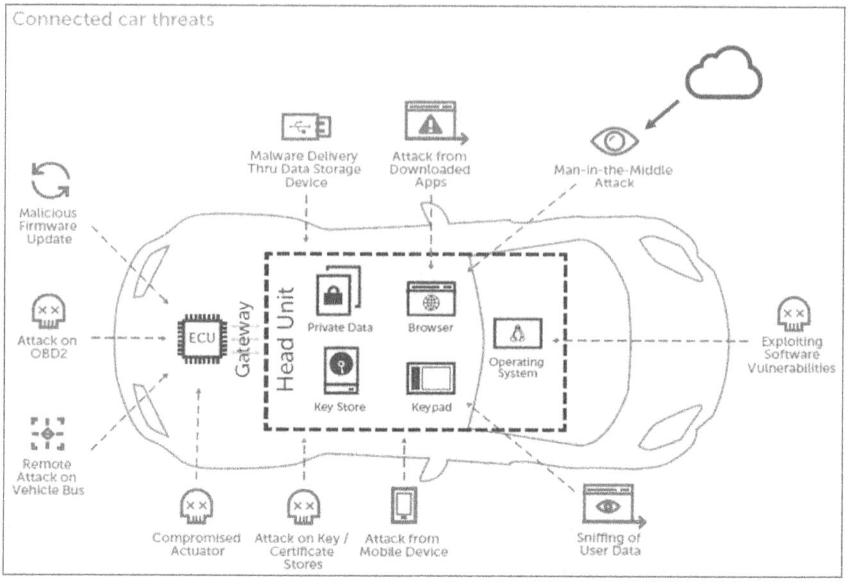

Figure 1. Description where it comes from (for external figures) url/publication and a date.

increased denial of customers to adopt CCAM, the loss of working hours, material damages, increased environmental pollution due e.g., to traffic jams or malicious modifications in sensors' firmware, and ultimately, the great danger for human lives, either they are drivers, passengers or pedestrians. CARAMEL's goal was to proactively address modern vehicle cybersecurity challenges applying advanced Artificial Intelligence (AI) and Machine Learning (ML) techniques, and also to continuously seek methods to mitigate associated safety risks. In order to address cybersecurity considerations for the already here autonomous and connected vehicles, well established methodologies coming from the ICT sector will be adopted, allowing to assess vulnerabilities and potential cyberattack impacts. Although past initiatives and cybersecurity projects related to the automotive industry have reached to security assurance frameworks for networked vehicles, several newly introduced technological dimensions like 5G, autopilots, and smart charging of Electric Vehicles (EVs) introduce cybersecurity gaps, not addressed satisfactorily yet.

The project has globally achieved all of its challenges, objectives, and milestones. The progress and the partner's commitment ensured the project's successful use cases implementation, despite the added complexity of deploying scenarios in real environments.

In the development activities the main goal was to develop the main components of an Advanced Cybersecurity Automotive system for each one of the three pillars of CARAMEL. This goal was pursued through the accomplishment of several activities, going from automotive threat modeling to the design and implementation of a solution for autonomous automated vehicles, plug-in electrical vehicles, and cooperative vehicles (V2X). For this last one, the V2X secure hardware and software were designed and developed considering the specific use case applications initially defined and launched to a demonstration within the , with a final demonstration event held in PASEU premises in Langen (Hessen, Germany) in June 2022 with satisfactory results and feedback.

Around research activities the main outcome was the exploration of how the recent advances in Artificial intelligence, Machine Learning, and other related technologies can help to improve cybersecurity awareness and defense against cyberattacks in the automotive sector. The work package's main tasks successfully ended during the first period of the project. WP5: The main outcome is to provide the design, development, and prototype implementation of a CARAMEL anti hacking device and in-depth defense solution. It is based on different cyber-security solutions to detect and mitigate cyber threats, with the processing and collection of large volumes of data in future autonomous vehicle scenarios. All these procedures have been executed with the best and most appropriate capabilities, suited to their functionalities, which have been adjusted in real time depending on the situational awareness about the underlying system at any time. WP6: The main outcome of WP6 is the integration of different parts of the prototypes developed in previous technical WPs into a consolidated and complete solution, tested and verified against specific test cases and KPIs. WP6 uses specific attack scenarios to perform an overall evaluation of the CARAMEL solution. WP6 work concluded with an overall assessment of CARAMEL as well as a roadmap for future evolution. WP7: During the project, the CARAMEL consortium participated in face-to-face or online events/conferences where different outputs were presented. Based on a description of the technical and organizational measures a safeguard action plan has been implemented to protect the rights and freedoms of the data subjects/research participants. For each deliverable and before submission to the EC we previously checked ethical requirements with the ethical committee before the Security Advisory Board review. (SAB)

The CARAMEL project strives to apply cybersecurity methodologies to the detection and mitigation of cybersecurity threats in the automotive domain. Innovation action in the project focuses on four selected pillars:

In pillar 1 (Autonomous Mobility) the project developed innovative technologies to detect attacks against the sensors in the vehicle, e.g. attacks using generative adversarial networks to disturb the object detection algorithms in the car. To this end, a novel intrusion detection system – the anti-hacking device – was integrated directly into the vehicle. The anti-hacking device uses machine-learning technology to detect attacks. Due to its unique design it will be easy and safe to

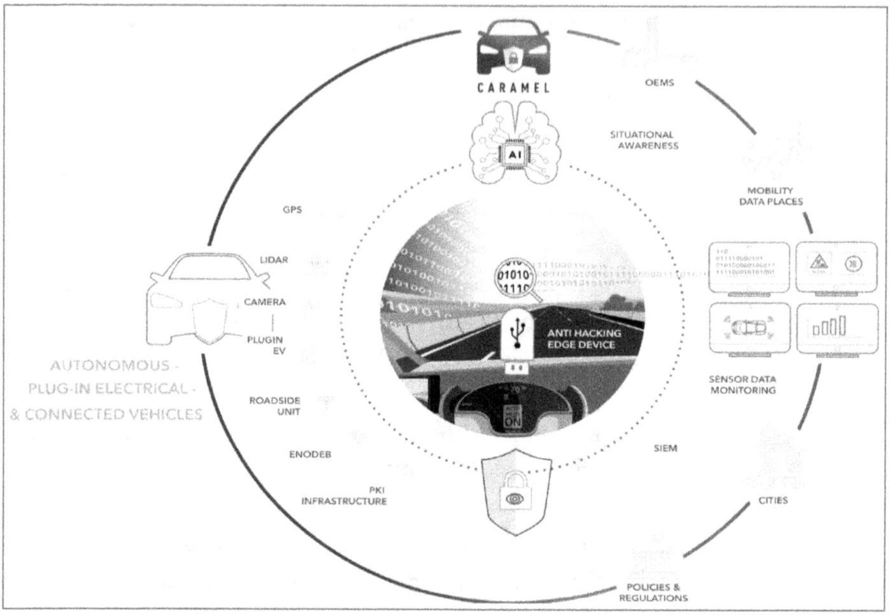

update these detection algorithms regularly to counter novel attacks in a short time. This research made autonomous or semi-autonomous driving more secure against advanced attack scenarios and will drive the acceptance and adoption of these innovations by the general public in Europe.

In pillar 2 (Connected Mobility) the project developed advanced attack detection technologies for the connected vehicle, especially the V2X protocols between the car and other cars and roadside units. Additionally, partners implemented an innovative threat detection mechanism for direct integration into on-board units (controllers built into the car). In a further development, the project developed an innovative bridge technology to make competing 802.11p and C-V2X networking technologies compatible using MEC devices or road-side units.

In pillar 3 (Electromobility) partners implemented security mechanisms to protect wide-scale attacks against the European EV charging infrastructure: Charging stations are part of the European critical infrastructure, attacks that cripple this infrastructure could adversely affect the flow of passengers and goods in Europe, the project implemented machine learning models to detect attacks on the level of the whole eCharging backend infrastructure so that early mitigations can be implemented.

In pillar 4 (Remote Control Vehicle) the Korean partners, who still have one year of project based on Korean funding, continued working in the implementation of remote control vehicle based on mmWAVE (23 GHz) use case, building a data processing architecture and developing a Malicious Traffic Detection Solution LSTM (Long-Short Term Memory)-based cyber-attack anomaly prediction/detection.

All these CARAMEL activities made the evolution of Automotive technology to a more connected and software-driven future more secure, therefore driving public acceptance of these technologies in Europe as well as giving European companies a competitive advantage.

Anti Hacking Device Concept/Vision

The CARAMEL anti-hacking device is designed as a passive intrusion detection device that is integrated as an additional controller into the vehicle (see Figure 2). The anti-hacking device *passively* listens to the car's internal busses and systems, processes and aggregates raw data from sensors and communication controllers and uses machine learning (ML) and other heuristics to detect possible attacks against the vehicle's systems.

It then *actively* creates attack reports (events) and sends them to the CARAMEL backend. Details of the integration of the anti-hacking device into the different CARAMEL scenarios are described in the CARAMEL specification.

The anti-hacking needs to be updated very frequently to run updated attack detection algorithms to counter newly discovered attack vectors. This requires frequent updates of the anti-hacking device firmware and application load. From a vehicle safety perspective any corruption of the anti-hacking device by bad actors must be avoided at all costs.

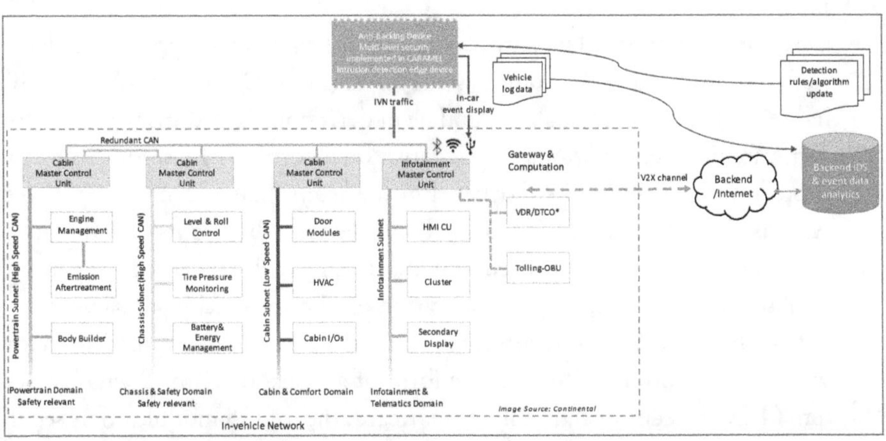

Figure 2. The anti-hacking device in the vehicle.

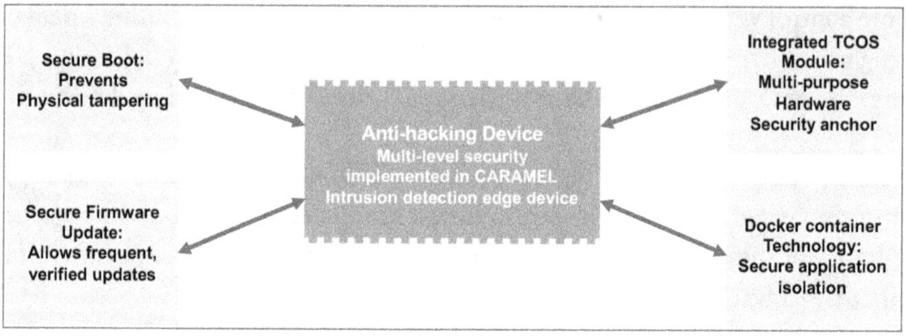

Figure 3. Anti-hacking device security features.

To this end multiple security features have been implemented in the project to harden the anti-hacking software against any kind of attacks (see Figure 3):

- **Secure Boot:** The anti-hacking device hardware has fuses (write-once programmable storage locations) that contain the public keys of acceptable boot loader signatures. The anti-hacking device only loads a correctly signed bootloader. The bootloader in turn verifies the signature of the Linux Kernel and only continues to load a verified kernel. These measures counter any physical tampering attacks on the boot medium.
- **Secure Firmware Update:** The anti-hacking device allows updating the firmware of the Internet (eg. over the vehicle's communication controller via LTE/5G). The anti-hacking device only accepts firmware update files that are properly signed by the anti-hacking device vendor. This protects the device against the installation of manipulated firmware images. In addition to this signature check the anti-hacking device implements also Secure Boot and would reboot to the last known safe state even if the secure firmware signature check were circumvented – effectively implementing multi-level security here.
- **Docker technology:** The anti-hacking device encapsulates the actual detection algorithms and also some system services into Docker containers. This has several advantages: It allows update of detection algorithms without a full firmware update. Additionally, the detection algorithms are separated by the protections offered by the Docker runtime against any mutual interference. As a last security measure the anti-hacking device only accepts signed Docker images from pre-defined trusted sources, effectively also implementing multi-level security for Docker implementation on the anti-hacking device.
- **Integrated TCOS (HSM) module:** The anti-hacking device contains a hardware secure module (HSM) in the form of a Telekom Card Operating System (TCOS) security chip. Like a smartcard, the TCOS module offers secure

storage of private key materials and certificates and the ability to run sensitive cryptographic operations securely on chip. The TCOS module offer these functionalities to Dockerized applications via a high-level security service also implemented as a Docker container.

Backend Solutions

Autonomous driving is a complex system in which the technologies associated with its operation require security systems aiming to mitigate cyber-attacks on the decision-making system. Although the detection of cyber-attacks allows warning the user about anomalies in their environment, a collective information system gathering threats detected by different autonomous systems would strengthen road safety by enabling the analysis of the information received and warn other systems about the anomalies identified in a given environment.

Not all vehicles would be prone to cyber-attacks to the same extent, because each system's configurations may differ, compromising threat detection. Furthermore, it is desirable to cross-validate the events witnessed from multiple sources and assign a level of confidence for some threats.

The use of a security operation center (SOC) with 2-way communication to the autonomous systems would add an extra layer of security that would only be achieved by enabling appropriate information exchange among vehicles.

The CARAMEL project developed a showcase system that demonstrates the benefits of bidirectional information exchange with vehicles via a cloud-based analytics backend to inform about detected user-relevant threats and warn other vehicles.

The CARAMEL backend receives messages from the vehicles, anti-hacking devices, MEC, etc. over an Information Exchange Protocol as threat messages alerting the vehicle's identified threat. All threat messages from vehicles, Anti-Hacking devices, MECs, etc. are gathered by the backend, saved in the database, and acknowledged to the sender. The backend further processes the threats that have been saved in the database to produce the "warning messages," which can be requested to be shared with other cars or requested by vehicles that already have a predefined route.

As part of the backend services, CARAMEL developed a threat visualization tool that allows it to monitor and investigate the threats acquired from each source. It also enables visualization of the severity of the detected threats. The project integrated threat detection such as denial of service, certificate revocation, V2X attacks, tampering of traffic signals, GPS spoofing, among others (Figure 4).

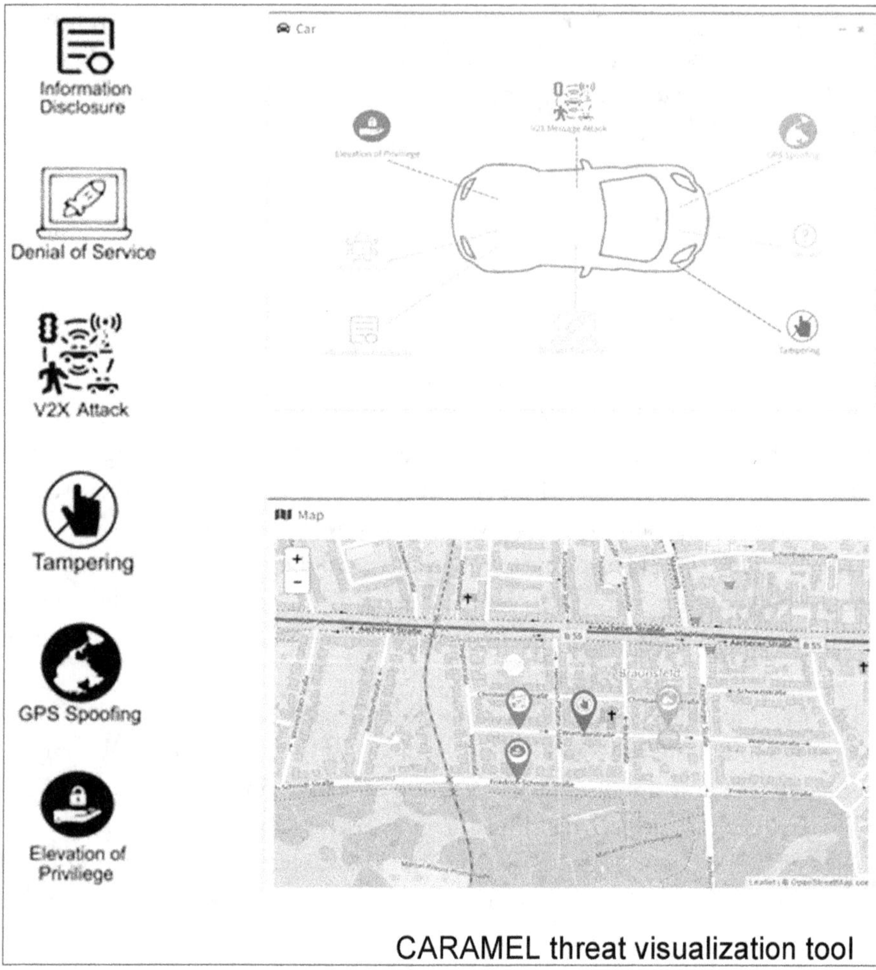

Figure 4. CARAMEL threat visualization tool.

DOI: 10.1561/9781638280613.ch1

Section 1

Autonomous Mobility

Cyberattacks do not require physical access to the vehicle or tampering with the communication system.

Introduction

Automated driving systems were developed to automate, adapt and enhance vehicle systems for safety and improved driving. Most road accidents occur due to human error, and automated systems use input from sensors like video cameras to reduce human error by issuing driver alerts or controlling the vehicle. Such systems have become common in modern cars, with automobile manufacturers integrating these systems in their cars. There are six levels of automation as shown in Figure 1. When it comes to Advanced Driver Assistance Systems (ADAS), the highest level (5) corresponds to full automation where the automated functions control all aspects of the car, and the lowest level (0) where the driver controls all aspects of the car.

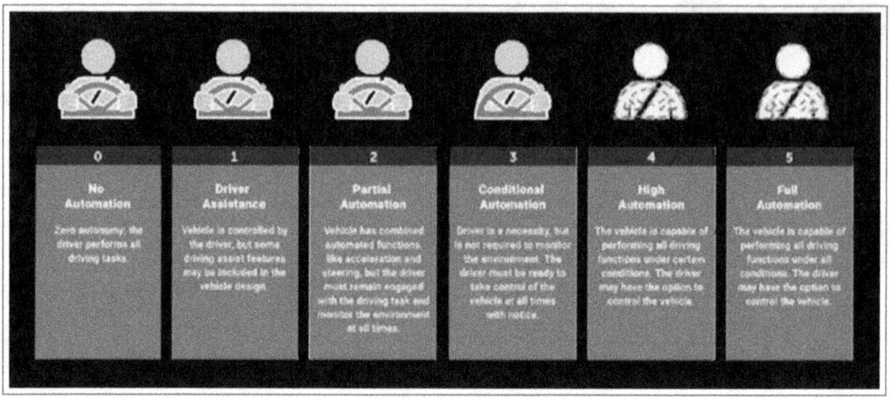

Figure 1. Society of automotive engineers (SAE) automation levels.

State of Art/Innovation

Recently, these systems have attracted increased attention within academia, and the academic community has begun to investigate the systems' robustness to various attacks. Recent studies [1–3] showed that ADAS alerts and notifications can be spoofed by applying adversarial machine learning techniques to scene structural elements (e.g. traffic signs, objects, etc.)

Adversarial attacks seek small perturbations of the input causing large errors in the estimation by the perception modality. Attacking perception functions using adversarial examples is a popular way to examine the reliability of learning approaches for data classification [4]. The key to all such attacks is that the change to the image should be minor yet have a large influence on the output. Adversarial examples typically involve small perturbations to the image that are not noticeable by the human eye. The adversaries are shown to work even when a single pixel is perturbed in the image [1]. Although these attacks reveal limitations of deep networks, they are hardly replicated in real-world settings. For instance, it is rather difficult to change a scene such that one pixel captured by a camera is perturbed in a specific way to fool the network. However, recent work on [5, 6] demonstrate that adversarial examples can also work when printed out and shown to the network under different illumination conditions. [7, 8] shows that adversarial examples can be 3D printed and are misclassified by networks at different scales and orientations. [8] constructs adversarial glasses to fool facial recognition systems.

[4, 9] show that stop signs can be misclassified by placing various stickers on top of them. Apart from the adversarial attacks, which involve scene modifications on the physical layer, within the Autonomous driving, the vulnerability of the Perception Engine is also an important issue to address. CARAMEL focuses on this point thanks to the perception engine contributed by Panasonic Automotive Europe.

Threats Considered/Detected

The Perception engine has to be secured against a variety of cyber-attacks at the sensors layer with the help of proper approaches for detecting the attacks and mitigating them. In line with what is described above, CARAMEL in the framework of autonomous mobility (pillar 1) believes the following scenarios presented in Table 1 are the most important cases to be addressed. Note that the presented scenarios are selected based on the CARAMEL consortium knowledge, available resources and showcasing capacity.

Scenario Description

CARAMEL scenarios on Autonomous Mobility can be classified into two big categories: physical adversarial attacks and attacks on the camera sensor. In this section, we will present them in more details. When referring to physical adversarial attacks we will consider attack scenarios where changes in the physical world will cause the cyber system in the autonomous vehicle to misbehave. Such is the example of physically manipulating traffic signs. Camera sensor attacks refer to the scenario where an attacker manages to access critical vehicle systems and manipulate directly the camera image. In such scenarios, detection and mitigation techniques will also utilize multiple additional sensor inputs such as lidar.

Caramel Engine Description (Solution Design)

Physical Adversarial Attacks

Deep learning solutions are used in several autonomous vehicle subsystems to perform perception, sensor fusion, scene analysis, and path planning. State-of-the-art and human-competitive performance have been achieved by ML on many computer vision tasks related to autonomous vehicles [10]. Nevertheless, over the last years it was demonstrated that ML solutions are vulnerable to certain visual attacks [11] that can cause the autonomous vehicles to misbehave in unexpected and potentially dangerous ways, for example on physical modification of the environment and especially traffic signs [11, 13].

It is considered that these attacks and modifications are physically added to the objects themselves. The traffic signs were selected as the main target domain of this scenario for several reasons discussed below:

- The relative visual simplicity of road signs.

Table 1. List of CARAMEL scenarios on autonomous mobility.

	Description
1	Adversarial attack on traffic signs: This is an attack on the physical layer. It assumes disturbance of the visual appearance of structural elements of the scene like the traffic signs. According to this attack, minor changes might be introduced, e.g.: stickers attached on the traffic signs in such a way that they might be marginally observable by the human eye but disturbing the scene perception output. The cyber-attack detection & mitigation engine will get as input the traffic sign topology from the perception engine and will assess the occurrence of cyber-attack.
2	Adversarial attack on lane/parking markings: As the above, this is also an attack on the physical layer. It is oriented towards distorting the appearance of lane/parking markings. The change could involve distortions in a multitude of appearance characteristics, e.g.: shape/length/colour. This attack should introduce minor changes in such a way that they could be marginally detectable by the human eye but finally affecting the output of the scene perception engine. The cyber-attack detection & mitigation engine should detect the occurrence of the cyber-attack and perform restoration in case that the restored version is derived with high confidence.
3	Attack on the Camera Sensor Layer: This scenario would involve a cyber-attack based on activating some malicious software which got installed during the software update process. Throughout this use-case the camera sensor could be attacked in a number of different ways, which could vary between adding noise lying on specific bands of the frequency spectrum/introducing morphological deformations/on the whole or parts of the image.
4	Attack on the Camera Sensor Layer by de-synchronizing the data: Throughout this scenario, the cyber-attack will be geared towards disturbing the association between the captured frames and the timestamp assigned to them. This will cause the failure of the perception engine, as all the architectural modules performing stochastic filtering on the scene observations will be affected by error. This use case should study the potential and the limitations of the cyber-attack detection and mitigation engine in assessing and recovering the failures.
5	Attack on the Camera Sensor by a remote agent: In addition to the aforementioned scenario, the cyber-attack detection and mitigation engine will be used to detect and mitigate the camera signal distortion in the case that a malicious remote agent interferes with the test vehicle by knowing the IP of the processing unit and sharing some erroneous data. More specifically, this use case will assume that the remote agent sends via V2X communication: time zone/daylight related data in order some sensor parameters (e.g.: gain/exposure time) to be tuned accordingly.
6	Attack on the LiDAR sensor: Apart from the camera, cyber-attacks on the LiDAR sensor is another important issue. As in use cases 3–5, the attack will involve triggering malicious software through either a remote agent or some date-related software update process. The malicious software could distort multiple attributes of the LiDAR signal which could vary by either adding noise to the measured data or changing arbitrarily some of the sensor configuration parameters (e.g.: scanning frequency).

Figure 2. Appearance perturbations on traffic signs.

- Road signs exist in a noisy unconstrained environment with changing physical conditions such as the weather, lighting, distance, and angle of the viewing camera,
- Road signs play an important role in transportation safety.
    - A reasonable threat model for transportation is that an attacker might not have control over a vehicle's systems but is able to modify the objects in the physical world that a vehicle might depend on to make crucial safety decisions.

In this scenario, the autonomous vehicle is expected to drive from a starting location to a given destination following a specified path. Throughout this path, certain traffic signs will be physically modified. An example could be the stop or turn left/right signs due to their important role in transportation safety. Figure 2 demonstrates an example of a physical attack from a real graffiti at the left and from an engineered attack aiming to make the ML system fail but most humans would not consider it suspicious [14].

Figure 3 represents the related scenario. The attacker modifies physical traffic signs. The autonomous vehicles with all the available sensors and mainly the camera aims to detect the attacks, provide notifications to the operator through HMI and to the other connected vehicles. Furthermore, the improved robust ML deep learning models should be able to overcome and not be affected by these attacks

The components existing in the traffic sign physical attack scenario are as follows:

- **Autonomous vehicle**: An autonomous vehicle embodying a multitude of on-board sensors (cameras, ultrasonic, GPS, Lidar, radar) and AI providing sufficient information related to the vehicle localization, the surrounding obstacles and the possibility of collision.

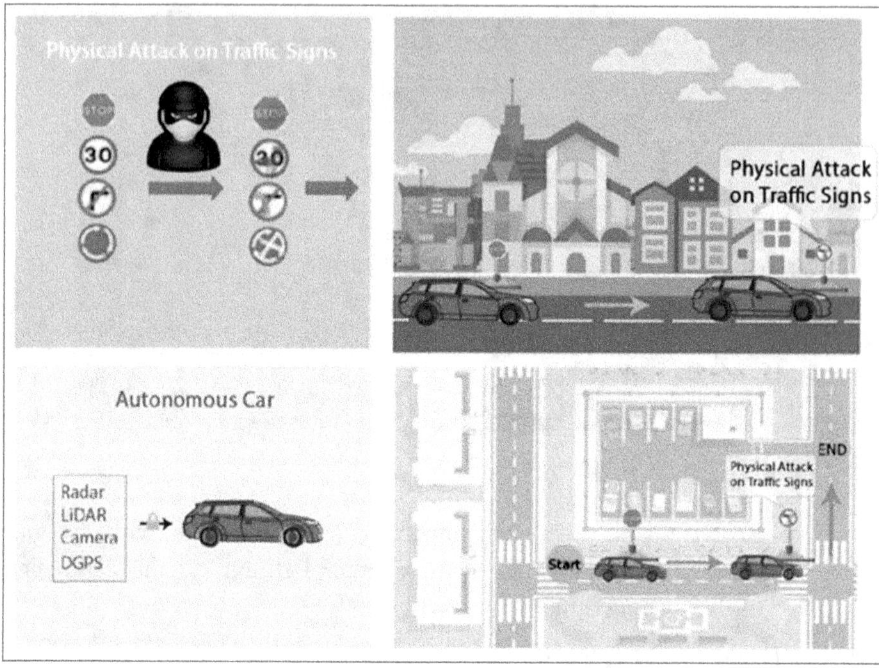

Figure 3. Adversarial attacks on traffic sign.

- **Visual sensors**: all the available vision related sensors (e.g. Camera, LiDAR, etc.) will be considered in this scenario. In practice the camera sensor will be utilized, as it is the main sensor for scene understanding and traffic sign detection and recognition.
- **Path planning system (PPS)**: PPS is a basic framework which defines the objective for the autonomous vehicle to move from one place to another. To achieve this, the PPS must choose a path and adjust to obstacles, terrain, and changing conditions to reach its destination safely. Note: The use case does not require/have access to the autonomous vehicle PPS to work.
- **Attacked traffic signs**: Real traffic signs will be modified and placed at the test area. Regarding the simulation models of traffic signs with attacks, they will be placed in the virtual environment.

CARAMEL components, which were integrated and the required functionalities from them are listed below:

- **Machine Learning component for sign attack detection (anomaly)**: ML models trained to detect attacks on traffic signs will be integrated to this scenario. The models will be based on various state-of-the-art architectures to detect anomalies.
- **Robust ML model for sign attached**: ML models trained to overcome such attacks will be integrated into the architecture.

Although adversarial attacks on the traffic signs comprise a very interesting use case for evaluating the potential and the limitations of CARAMEL's solution on addressing cyber-attacks, the possibility of reproducing this use case on the test area with Panasonic's vehicle remains to be verified based on the input of the test area operator. Given the fact that the test area is managed by a third-party service provider, the consensus of the operator on distorting the appearance of existing traffic sign structures needs to be provided. However, this scenario will be extensively investigated in the simulator as the flexibility provided there by the simulation environment will allow detailed analysis on the precision of ML module in detecting and mitigating the attacks.

Table 2 describes the traffic sign attack scenario while Figure 4 shows the roles of the actors identified for this use case.

- **Vehicle operator/passenger –** a person responsible to operate the vehicle in the case of a not fully automated one, monitoring the environment and the vehicle behaviour. They are responsible for receiving notifications from the CARAMEL platform and taking the necessary measures to react to the physical attacks.
- **Connected Vehicles –** a list of other vehicles connected to the current one and the corresponding operators or passengers. They are responsible for receiving related notifications and acting accordingly.

Table 2. Traffic sign attack scenario definition.

Use Case	Scenario ID and Title	Priority Level
Autonomous Vehicles – Traffic Sign Physical Attack	Detection and reaction to physical attacks on traffic signs	High
	Robustness to physical attacks on traffic signs	High

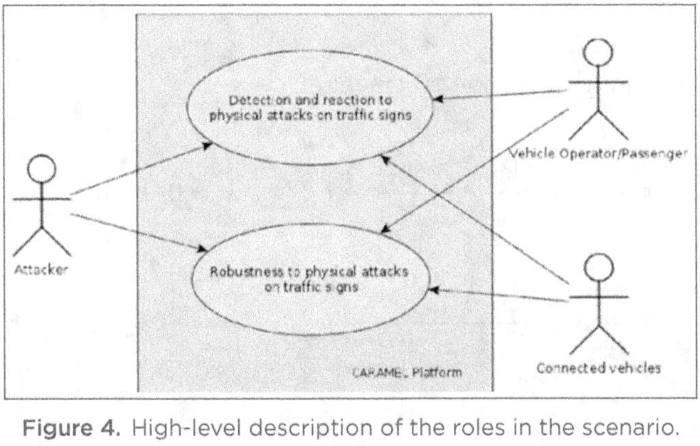

Figure 4. High-level description of the roles in the scenario.

- **Cyber-attacker** – a person conducting the physical-attack either randomly or considering adversarial permutations on physical objects such as traffic signs.

As discussed in Table 3 apart from the adversarial attacks geared towards physically distorting the appearance of some of the scene structural elements, CARAMEL also targets to study the potential of cyber-attack detection techniques in estimating the occurrence of attacks on the sensor signal, camera being the most possible candidate.

GPS Spoofing Cooperative

Consider a 2-D region where N connected vehicles of a VANET collect and exchange measurements while moving. An example of such a VANET, is shown in Figure 5. The location of the i-th vehicle at k-th time instant is given by $x_i^{(k)} = [x_i^{(k)} \, y_i^{(k)}]^T$.

Each vehicle knows its absolute position from GPS and measures its relative distances and angles with respect to neighbouring vehicles using LIDAR or RADAR. The true relative distance $z_{d,ij}^{(k)}$ between connected vehicles i and j is given by $z_{d,ij}^{(k)} = \|x_i^{(k)} - x_j^{(k)}\|$, where $\| \cdot \|$ is the l^2 norm. The true angle $z_{a,ij}^{(k)}$ (shown in Figure 6) between neighbouring vehicles i and j is given by $z_{a,ij}^{(k)} = arctan\frac{y_j^{(k)} - y_i^{(k)}}{x_j^{(k)} - x_i^{(k)}}$.

The acquired measurements are assumed to be described by the following models:

- Relative distance measurement:

$$\tilde{z}_{d,ij}^{(k)} = z_{d,ij}^{(k)} + w_d^{(k)}, \quad w_d^{(k)} \sim N(0, \sigma_d^2) \tag{1}$$

- Relative angle measurement:

$$\tilde{z}_{a,ij}^{(k)} = z_{a,ij}^{(k)} + w_a^{(k)}, \quad w_\alpha^{(k)} \sim N(0, \sigma_a^2) \tag{2}$$

- Relative azimuth angle measurement:

$$\tilde{z}_{az,ij}^{(k)} = \lambda\pi + arctan\frac{|x_j^{(k)} - x_i^{(k)}|}{|y_j^{(k)} - y_i^{(k)}|} + w_{az}^{(k)}, \quad \lambda = 0, 1 \quad \text{or}$$

$$\tilde{z}_{az,ij}^{(k)} = \lambda\pi + arctan\frac{|y_j^{(k)} - y_i^{(k)}|}{|x_j^{(k)} - x_i^{(k)}|} + w_{az}^{(k)}, \quad \lambda = \frac{1}{2}, \frac{3}{2}, \quad w_{az}^{(k)} \sim N(0, \sigma_{az}^2) \tag{3}$$

Table 3. Summarization the physical adversarial attack and identifies the evaluation criteria.

Scenario Name	Detection of Physical Attacks on Traffic Signs
Related Pillar	Autonomous Vehicle Simulator
Scenario Description	The scenario deals with two kinds of attack: attacker vandalizes traffic signs i.e. some random graffiti that hides a different part of the sign or a coordinated attack such as generating ML based image to cover the signs.
Brief Description	The autonomous vehicle moves in the test area. Certain traffic signs have been physically modified to influence the driving behaviour and planning of the autonomous vehicle. CARAMEL's platform is operating in parallel to the driving system of the autonomous vehicle without influencing the decision-making module. When the vision-related sensor and the ML components of CARAMEL detects a physical attack, a corresponding notification will be displayed to the vehicle operator or passenger.
Challenges	1. Ability to detect physical attacks on traffic signs. 2. Improved robustness on physical attacks.
Assumptions & Pre-Conditions	1. Datasets (real and synthetic) for traffic signs are available. 2. The camera and vision sensors are properly calibrated for both the real and simulated cases.
Goal (Successful End Condition)	The physical attack on the traffic signs is successfully identified without affecting the driving behaviour and the decision-making processes of the autonomous vehicle.
Involved Actors	1. Attacker. 2. Vehicle operator/passenger.
Scenario Initiation	An autonomous vehicle from a given location is instructed to drive to a selected end destination.
Main Flow	1. Selection of starting and ending location. 2. Data acquisition mainly from the camera sensor. 3. Physical attack detection for traffic signs. 4. Robust model deployment in parallel with step 3. 5. Operator notification.

(*Continued*)

Table 3. Continued

Scenario Name	Detection of Physical Attacks on Traffic Signs
Evaluation Criteria	CARAMEL platform detects the attacked signs and notifies the vehicle operator allowing them to take appropriate remedial actions. Metrics related to the detection and recognition accuracy such as *F1* score, Precision and Recall will be considered.

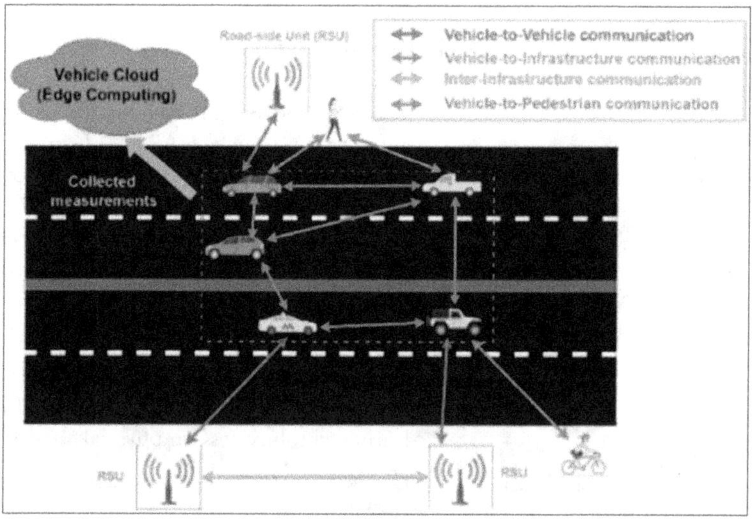

Figure 5. VANET.

- Absolute position measurement:

$$\tilde{z}_{p,i}^{(k)} = x_i^{(k)} + w_p^{(k)}, \quad w_p^{(k)} \sim N(0, \Sigma_p) \tag{4}$$

The difference between the angle and the azimuth angle measurement is depicted in Figure 6. Covariance matrix Σ_p is a diagonal matrix equal to $diag\ (\sigma_x^2, \sigma_y^2)$. A typical approach in the area of Cooperative Localization (CL) is to formulate an objective cost function $C(x)$ according to Maximum Likelihood Estimation (MLE) criterion and to minimize it with respect to locations x_i in order to reduce the error of absolute position measurement. The likelihood function of the measurement models can be written as:

$$L(x) = \prod_{i \in N, j \in N(i)} P(x_i^{(k)}, x_j^{(k)}) \prod_{i \in N, j \in N(i)} P(x_i^{(k)}, x_j^{(k)}) \prod_{i \in N} P(x_i^{(k)}), \tag{5}$$

where $N(i)$ denotes the set of neighbours of the i-th vehicle and $P(\cdot)$ are the probability density functions of the measurement models. If we take the logarithm of

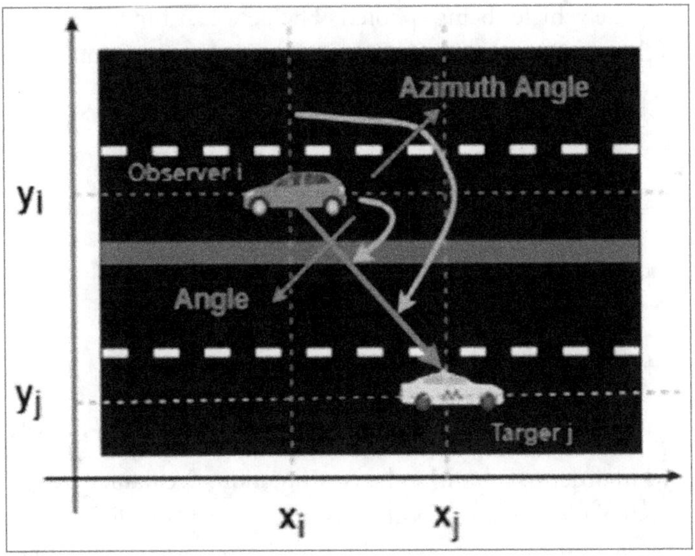

Figure 6. Angle and azimuth angle measurement.

Eq. (5), then the objective cost function (same as in [12] and similar to that of [13]) is given by:

$$C(x^k) = \sum_{i\in N, j\in N(i)} (\tilde{z}_{d,ij}^{(k)} - z_{d,ij}^{(k)})^2/2\sigma_d^2 + \sum_{i\in N, j\in N(i)} (\tilde{z}_{a,ij}^{(k)} - z_{a,ij}^{(k)})^2/2\sigma_a^2$$
$$+\sum_{i\in N} \frac{1}{2}[(\tilde{z}_{p,i}^{x,(k)} - x_i^{(k)})^2/\sigma_x^2 + (\tilde{z}_{p,i}^{y,(k)} - y_i^{(k)})^2/\sigma_y^2] \tag{6}$$

The GPS spoofing attack impacts on the absolute position measurement that is provided to vehicles. It may result in dozens, hundreds or even thousands of meters away from true location. Let $O_i^{(k)} = [o_i^{x,(k)}\ o_i^{y,(k)}] \in R^{N\times 2}$ be the unknown matrix of outliers to the true x and y locations of vehicles, which models the spoofing attack. Our main goal is to retrieve that impact and to substitute it from cooperative locations estimation approach. The spoofed absolute position measurement is now provided by the following model:

- Spoofed absolute position measurement:

$$\underline{z}_{p,i}^{(k)} = \tilde{z}_{p,i}^{(k)} + O_i^{(k)} \tag{7}$$

The main hypothesis of the robust cooperative localization solutions that will be developed, relies on the fact that only a small number of 20–25% of VANET's vehicles can be compromised. That property facilitates the exploitation of l^1 norm minimization approaches, since the outliers matrix is actually sparse, because it corresponds to the vehicles being spoofed. The new cost function, based on MLE criterion and the sparsity properties of outliers matrix, can be formulated according to Eq. (8):

$$
\begin{aligned}
C(x^k) = & \sum_{i\in N, j\in N(i)} (\tilde{z}_{d,ij}^{(k)} - z_{d,ij}^{(k)})^2/2\sigma_d^2 + \sum_{i\in N, j\in N(i)} (\tilde{z}_{a,ij}^{(k)} - z_{a,ij}^{(k)})^2/2\sigma_a^2 \\
& + \sum_{i\in N} \frac{1}{2}[(\underline{z}_{p,i}^{x,(k)} - o_i^{x,(k)} - x_i^{(k)})^2/\sigma_x^2 + (\underline{z}_{p,i}^{y,(k)} - o_i^{y,(k)} - y_i^{(k)})^2/\sigma_y^2 \\
& + \lambda_1 \| o^{x,(k)} \|_1 + \lambda_2 \| o^{y,(k)} \|_1,
\end{aligned} \tag{8}
$$

where $\| \cdot \|_1$ is the l^1 norm. The interior point methods provided by CVX software can be applied in order to minimize the cost function. We named this approach as Robust Traditional Cooperative Localization based on MLE (RTCL-MLE).

An alternative approach is to treat the VANET as an undirected graph, using the connected vehicles as its vertices and the communication links between them as its edges. The associated Extended Laplacian Matrix $\tilde{L}^{(k)}$ of the VANET graph and the differential coordinates $\delta_i{}^{(k)} = [\delta_i^{x,(k)}\ \delta_i^{y,(k)}] \in R^{N\times 2}$ of each vehicle, can be derived according to that graph modelling and the previously discussed measurement models. See [23, 24] for more details on the Laplacian Processing for Cooperative Localization. The differential coordinates are equal to:

$$
\delta_i^{x,(k)} = \frac{1}{d_i^{(k)}} \sum_{j\in N(i)} -\tilde{z}_{d,ij}^{(k)} sin\tilde{z}_{az,ij}^{(k)}
$$

$$
\delta_i^{y,(k)} = \frac{1}{d_i^{(k)}} \sum_{j\in N(i)} -\tilde{z}_{d,ij}^{(k)} cos\tilde{z}_{az,ij}^{(k)},
$$

where $d_i{}^{(k)}$ is the number of connected neighbors to i-th vehicle. Afterwards, the two following vectors are formed:

$$
b^{x,(k)} = [\delta^{x,(k)}\ \tilde{z}_p^{x,(k)}]^T \in R^{2N}
$$

$$
b^{y,(k)} = [\delta^{y,(k)}\ \tilde{z}_p^{y,(k)}]^T \in R^{2N}
$$

assuming that the noisy GPS positions of the vehicles of the network act as the anchors. Thus, the two following minimization problems have been formulated, based on the graph representation of VANET and the sparsity properties of outliers vectors, in order to estimate the locations of N vehicles, while in the same time to detect and mitigate possible attacks on GPS measurements:

$$\underset{\boldsymbol{x}^{(k)},\boldsymbol{o}^{x,(k)}}{\operatorname{argmin}} \|\tilde{\boldsymbol{L}}\boldsymbol{x}^{(k)} - (\boldsymbol{b}^{x,(k)} - \boldsymbol{q}^{x,(k)})\|^2 + \lambda_3\|\boldsymbol{o}^{x,(k)}\|_1$$

$$\underset{\boldsymbol{y}^{(k)},\boldsymbol{o}^{y,(k)}}{\operatorname{argmin}} \|\tilde{\boldsymbol{L}}\boldsymbol{y}^{(k)} - (\boldsymbol{b}^{y,(k)} - \boldsymbol{q}^{y,(k)})\|^2 + \lambda_4\|\boldsymbol{o}^{y,(k)}\|_1$$

Once again, the interior point methods provided by CVX software can be applied in order to solve the two minimization problems. Note that vectors $q^{x,(k)}$ and $q^{y,(k)}$ are equal to:

$$q^{x,(k)} = [0\ o^{x,(k)}]^T \in R^{2N}$$

$$q^{y,(k)} = [0\ o^{y,(k)}]^T \in R^{2N},$$

where zero vector $0 \in R^N$. The outliers of the position must be removed only from the anchors part of vectors. $b^{x,(k)}$, $b^{y,(k)}$. We named this approach as Robust Graph based Centralized Cooperative Localization (RCGCL). Note that in both cooperative robust methods, regularizing parameters $\lambda_{1,2,3,4} > 0$ control the minimization of location estimation term and the outliers estimation term.

During the detection phase of either of the two robust schemes, a vector containing the Euclidean distances between the initial GPS locations and the estimated locations is formed. Afterwards, a small threshold equal to 10 is set, implying that distances below 10 m do not correspond to attacked vehicles, while distances greater than 10 m may be indicative of an attack. In the latter case, k-means clustering algorithm, with k = 2, is applied on the corresponding distances, producing two clusters with associated centers. The cluster with the largest center contains, in fact, the distances that correspond to attacks. As such, the ids of spoofed vehicles can be identified.

Apparently, the collaboration towards multi-modal fusion among the vehicles of VANET can lead to estimating their locations more accurately than GPS, as well as defending against GPS spoofing attack. More specifically, exploiting the sparsity properties of outliers matrix, our defence mechanism mitigates the impact of spoofing and detects the compromised vehicles. The overall collaborating defence mechanism is depicted on Figure 7.

We have validated the collaborating and robust to GPS spoofing attack approaches in a simulated environment. A number of vehicles, say N = 20, constitute the VANET. For a reduced computational load, two vehicles communicate if and only if their distance is up to 20 m and the maximum number of connected neighbours is 6. We have chosen $\sigma_x = 3\ m$, $\sigma_y = 2.5\ m$, $\sigma_d = 1\ m$ and $\sigma_a = \sigma_{az} = 4°$. The true trajectories of the first 3 vehicles moving for 500 time

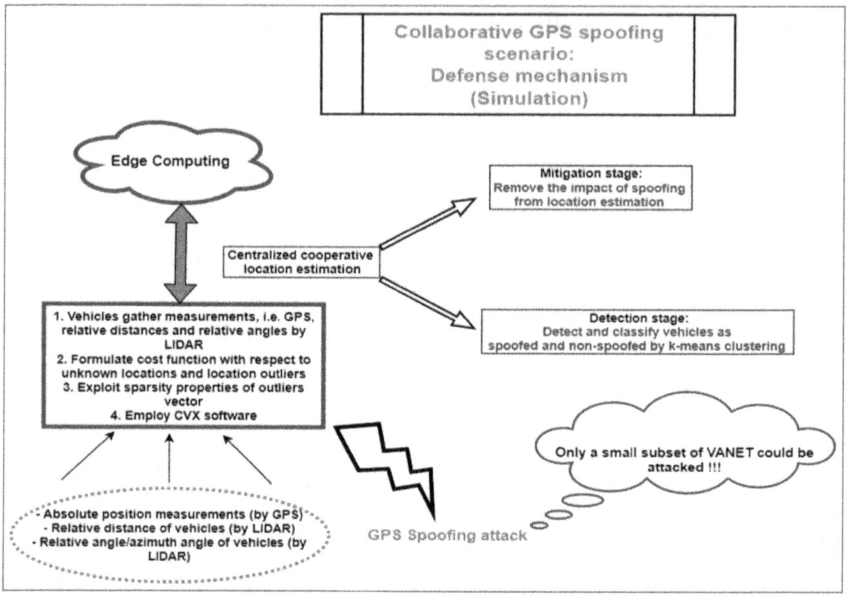

Figure 7. High-level architecture of collaborating defence mechanism.

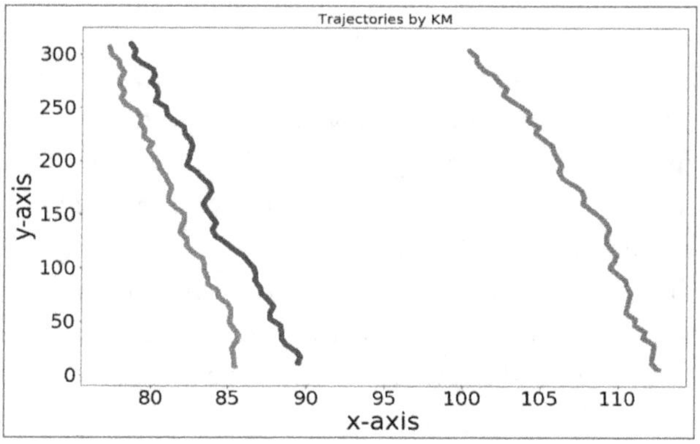

Figure 8. True trajectories of 3 vehicles.

instances are depicted on Figure 8. They have been created according to the bicycle kinematic model (KM). The spoofing attack is simulated by adding a bias (sampled uniformly in the interval of [5, 40]) to randomly chosen vehicles at each time instant, resulting in an average deviation of the true location equal to 34 m. The experiments were conducted for a number of 5%, 10%, 20% and 30% compromised vehicles. We constructed the Cumulative Distribution Function (CDF) of Localization Mean Square Error (LMSE) of RTCL-MLE, RCGCL, spoofed GPS and normal GPS without outliers. In Figure 9, the CDFs of LMSE for 1 (5%) 2 (10%), 4 (20%) and 6 (30%) compromised vehicles are being presented. It is clearly evident that both the robust schemes significantly reduced the error of

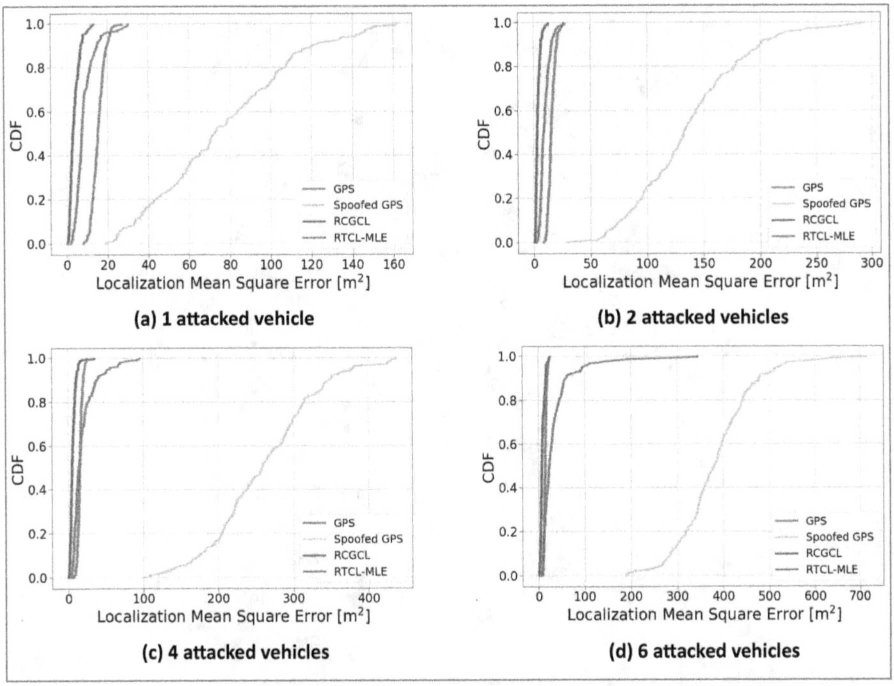

Figure 9. CDFs of LMSE of the proposed cooperative and robust schemes for different number of compromised vehicles.

spoofed GPS. Moreover, RGCL achieves much greater performance than RTCL-MLE and even the theoretically intact GPS. Based on that, the reduction of LMSE of RCGCL and RTCL-MLE with respect to intact GPS, was 77% and 56%, respectively, for 5% compromised vehicles. Moreover, the reduction of LMSE of RGCL and RTCL-MLE with respect to intact GPS, was 79% and 47%, respectively, for 10% compromised vehicles. However, for 20% compromised vehicles, LMSE was reduced by 65% with RCGCL, but increased by 1.02% with RTCL-MLE. Finally, for 30% compromised vehicles, LMSE was reduced by 53% with RCGCL, but increased by 2.9% with RTCL-MLE. As it was expected, when the number of attacked vehicles increased, the performances have been degraded. The two proposed cooperative approaches achieved significant reduction of spoofed GPS error, by estimating accurately the locations of vehicles. Furthermore, RCGCL proved to outperform RTCL-MLE.

We have also employed the CARLA autonomous driving simulator, in order to generate realistic urban trajectories. More specifically, we generated the trajectories of 200 vehicles moving for 200 time instances in a simulated city. Figure 10 presents the simulated city by CARLA. Once again, we randomly spoofed some vehicles by adding a bias to their GPS measurement. The main hypothesis is that the spoofed vehicles belong to a cluster of connected vehicles, otherwise our defence

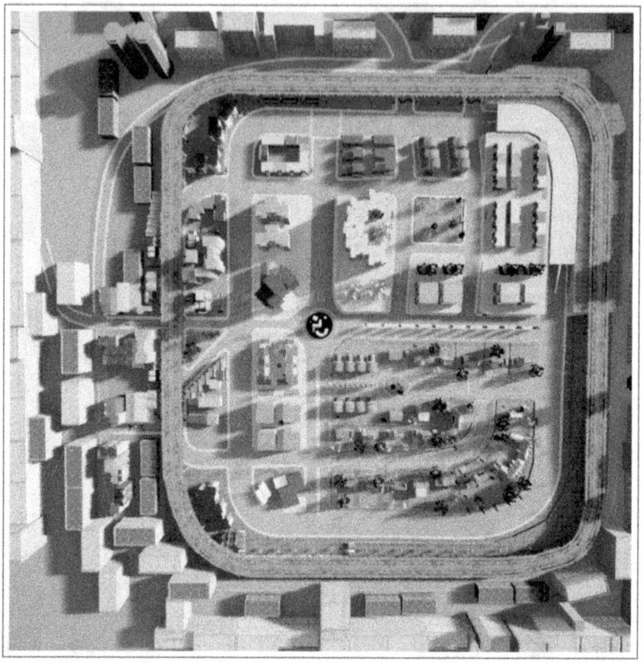

Figure 10. Simulated city by CARLA.

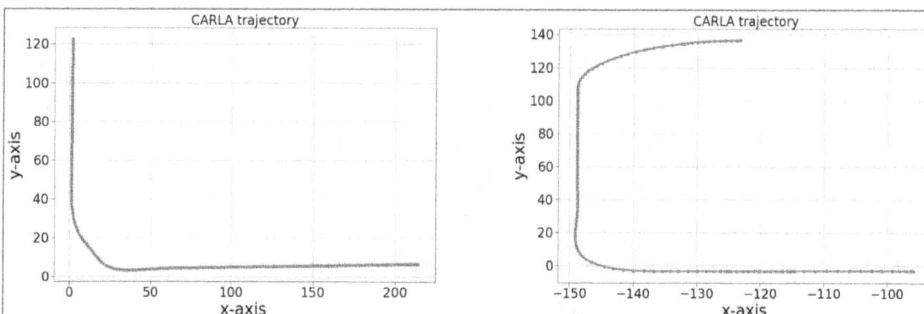

(a) CARLA trajectory of vehicle #128 **(b) CARLA trajectory of vehicle #195**

Figure 11. Trajectories of two CARLA vehicles.

mechanism is not applicable, while the number of compromised vehicles was set to 15% of the associated cluster's size. In Figure 11 the true trajectories by CARLA of two vehicles are depicted. In Figure 12, the CDFs of Localization Error for RTCL-MLE, RCGCL, spoofed GPS and normal GPS without outliers of the two compromised vehicles are presented. In Figure 12-(a) the reduction of Localization Error with respect to theoretically intact GPS is 62% with RCGCL, while it increases by 0.81% with RTCL-MLE. In Figure 12-(b) the reduction of Localization Error with respect to theoretically intact GPS is 61% with RCGCL, while it increases by 1.30% with RTCL-MLE. Although both robust schemes mitigate successfully

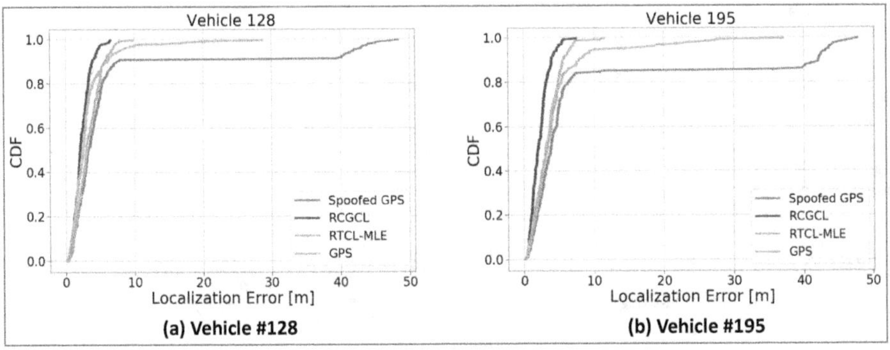

Figure 12. Localization Error of spoofed vehicles.

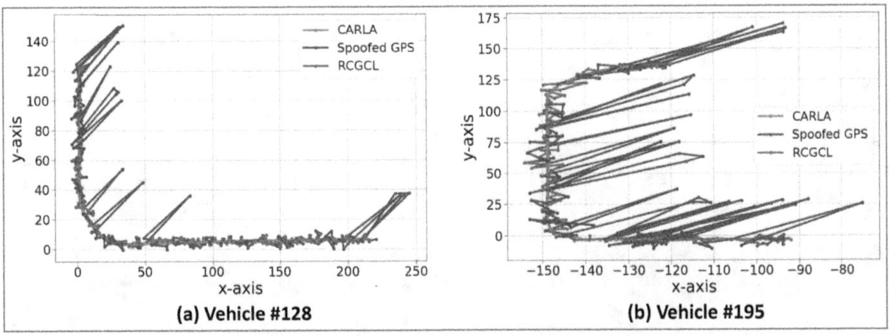

Figure 13. Reconstructed trajectories.

the effects of GPS spoofing, only RCGCL is able to achieve greater performance, even than the theoretically normal GPS without outliers. Figure 13 presents the trajectories by CARLA, spoofed GPS and the estimated by RCGCL. The collaborating scheme tackles with GPS spoofing, and is able to retrieve much more accurate locations.

During the detection stage, we measured True Positives, False Positives, True Negatives, False Negatives, True Positive Rate and False Positive Rate for the entire simulation horizon (500 time instances). Afterwards, we constructed the Receiver Operating Characteristics (ROC) curves for 5%, 10%, 20% and 30% spoofed vehicles and measured the Area Under Curve (AUC). The ROC curves for the two schemes are depicted on Figure 14. In Figure 14(a), RGCL and RTCL-MLE both achieved 99% AUC. In Figure 14(b), they achieved 95% AUC. In Figure 14(c), RGCL achieved 95% AUC, while RTCL-MLE 94% AUC. Finally, in Figure 14(d), they achieved 95% and 93%, respectively. Regardless the number of attacked vehicles, the two robust schemes were able to detect the spoofed vehicles, since the classification accuracy was very high, i.e. AUC greater than 90%. We notice also that as the number of compromised vehicles increases, the classification accuracy

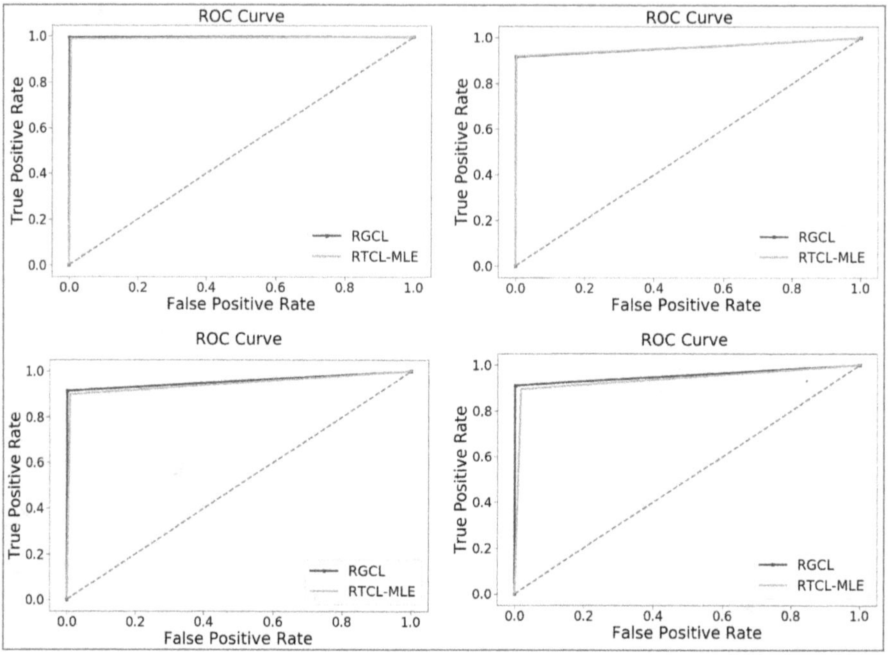

Figure 14. ROC curves for different number of compromised vehicles, (a) 5% attacked vehicles, (b) 10% attacked vehicles, (c) 20% attacked vehicles, (d) 30% attacked vehicles.

is slightly reduced. RGCL performs the same or even better than RTCL-MLE. However, due to its much better performance during the mitigation stage, RGCL proves its superiority as a collaborating defence mechanism against GPS spoofing.

Noise Attack at the Sensor Level

Several filtering methods based on nonlinear and nonstationary signal processing, such as Kalman filtering (KF), wavelet transform (WT), and empirical mode decomposition (EMD), have been proposed to denoise lidar signals and KF can be defined as a recursive estimation algorithm with minimum mean square error as the best criterion; however, this method will lose accuracy when the aerosol extinction coefficient changes sharply. WT can decompose the signal into different frequencies through multiscale analysis and has good time–frequency localization characteristics. However, WT is hampered by the problems of selecting suitable wavelet basis functions and the decomposition level [14]. EMD technology makes up for the limitations of WT and KF and has good adaptability, thereby enabling it to efficiently reflect the local frequency characteristics of the signal. Nevertheless, EMD and its variants still have some drawbacks, such as mode mixing and end effects [15]. Variational mode decomposition (VMD), proposed by Dragomiretskiy and Zosso in 2014, is a new adaptive signal decomposition method that has a different process of mode decomposition compared with EMD [16]. It has obvious advantages in dealing with non-recursive signals, which can not only overcome the mode mixing problems in EMD but also obtain a better filtering effect by using its own Wiener filtering characteristics. VMD has been successfully implemented in many fields, such as mechanical diagnosis, biomedical sciences, and hydropower unit vibration signal processing [15, 16, 17, 20]. Despite VMD's contributions, two crucial parameters need to be resolved: the decomposition mode number K and the quadratic penalty α [21]. These two parameters are usually selected in a certain range. If they are obtained by the trial-and-error method, it will require tens of thousands of operations and will waste a lot of time. For this reason, the parameter values are usually determined based on experience and convenience, which greatly limits the performance of the VMD method and may cause inaccurate decomposition results. Therefore, appropriate methods are needed to obtain the optimal values of these two parameters, and various algorithms have been proposed. Li et al. proposed an independence oriented VMD method, which finds the most suitable mode number by peak searching and the similarity principle; however, this method does not consider the influence of the bandwidth control parameter on decomposition results [21]. Shi et al. investigated a precise feature extraction method that optimizes the two parameters K and α for VMD independently [18]. However, they neglected the interaction between the two parameters [19, 20]. Selecting relevant modes is also an important issue. At present, the indicators for selecting relevant modalities are correlation coefficient, permutation entropy, approximate entropy, and Hausdorff distance, among others [19–22].

Image denoising techniques have attracted much attention in recent 50 years. At the outset, nonlinear and non-adaptive filters were used for image applications [21]. Nonlinear filters can preserve the edge information to suppress the noise, unlike linear filters [23]. Adaptive nonlinear filters depend on local signal-to-noise ratios to derive an appropriate weighting factor for removing noise from an image corrupted by the combination of additive random, signal dependent, impulse noise and additive random noise [23]. Non-adaptive filters can simultaneously use edge information and signal-to-noise ratio information to estimate the noise. In time, machine learning methods, such as sparse-based methods were successfully applied in image denoising [24]. A non-locally centralized sparse representation (NCSR) method used nonlocal self-similarity to optimize the sparse method and obtained high performance for image denoising [25]. To reduce computational costs, a dictionary learning method was used to quickly filter the noise [26]. To recover the detailed information of the latent clean image, priori knowledge (i.e., total variation regularization) can smooth the noisy image in order to deal with the corrupted image [27, 28]. More competitive methods for image denoising including the Markov random field (MRF)[28], the weighted nuclear norm minimization

(WNNM), learned simultaneous sparse coding (LSSC) [29], cascade of shrinkage fields (CSF) [27], trainable nonlinear reaction diffusion (TNRD) and gradient histogram estimation and preservation (GHEP) [29].

Adversarial Attack at the Scene Understanding Level

To segment, detect, and classify objects in an autonomous vehicle scene with robust and discriminative performance there are quite a few challenges that need to be addressed. Despite current state-of-art methods based solely on camera data seem to achieve astonishing results under normal imaging conditions, they fail in adverse weather and imaging conditions. Existing training datasets are biased towards clear weather conditions, and detector architectures are designed to rely only on the redundant information in the undistorted sensory streams. So, in a scenario that a sensor fails on specific conditions or, as in our case, has been attacked using a system relying on multiple sensor modalities gives a more robust result. There are three different fusion strategies that have been used so far in literature in order to exploit the advantages that each modality offers. So, we have early, late and deep fusion:

- **Early fusion:** Modalities are combined at the beginning of the process, creating a new representation that is dependent on all modalities.
- **Late fusion:** Modalities are processed separately and independently up to the last stage, where fusion occurs. This scheme does not require all modalities to be available as it can rely on the predictions of a single modality.
- **Deep fusion:** Modalities are mixed hierarchically in neural network layers, allowing the features from different modalities to interact over layers, resulting in a more general fusion scheme.

We choose to go with a late fusion strategy. Given that the camera is attacked the data coming from it would be unreliable to use them in the feature extraction level. In late fusion, or else decision fusion, multiple classifiers are used to generate decisions that are then combined to form a final decision, as shown in Figure 15.

Specifically, in our case, we fuse the output of our segmentation model with the output of a deep learning model for 3D object detection based only on raw LiDAR frames. So, we can decide whether the camera has been attacked or not by correlating the two outputs. The overall architecture can be seen in Figure 16. Next, we will analyse the structure of the deep learning model that has been used for object detection.

Current state-of-art methods for 3D object detection proposed several ways to extract feature information from the sparse 3D point clouds. Many researchers tried to exploit pre-existing 2D CNN architectures by projecting point cloud to bird's view for 3d box generation [11, 12, 13]. Another approach that has quite reliable

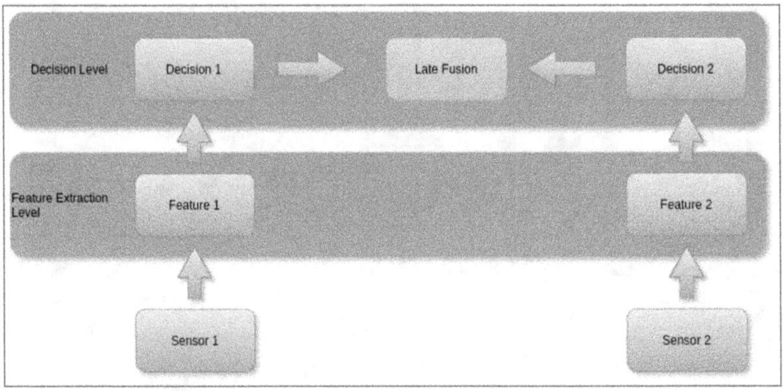

Figure 15. Abstract sensor fusion architecture describes where the late fusion occurs.

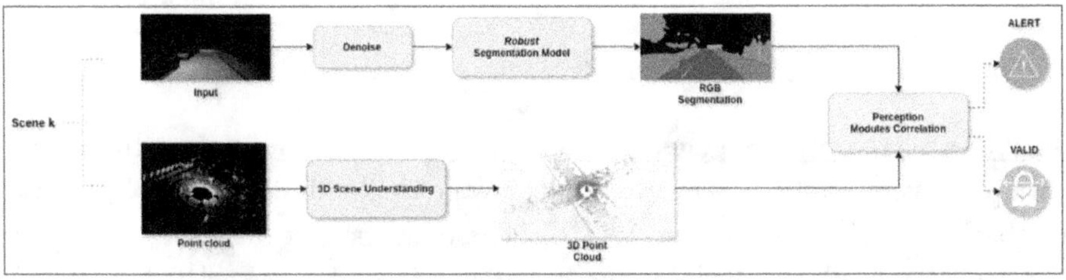

Figure 16. The overall pipeline of the perception module.

results is to group the points into voxels and with the use of 3D CNN to learn the features of voxels to generate 3D boxes [14, 15]. Due to the projection, and the voxelization stage the aforementioned methods suffer from loss of information. A significant step towards a better scene analysis of 3d raw point clouds was the network PointNet and PointNet++ [16, 17] PointNet architecture directly learn point features from raw point clouds, which greatly increases the speed and accuracies of point cloud classification and segmentation, instead of representing the point cloud as voxels or multi-view formats. The network PointRCNN [18], that we used for object detection uses PointNet as a backbone. More details about PointRCNN [18], are provided in the next section.

PointRCNN Architecture for Point Cloud 3D Detection

The model that we used is PointRCNN [18] and is one of the current state-of-art models for 3D object detection. The overall architecture of the model as it is proposed in [18] is illustrated in Figure 17. PointRCNN [18] is a bottom-up point cloud-based 3D bounding box proposal generation algorithm, which generates a small number of high-quality 3D proposals via segmenting the point cloud into foreground objects and background. So, the model consists of a bottom-up 3D proposal generation stage and a stage for the refinement of the bounding boxes.

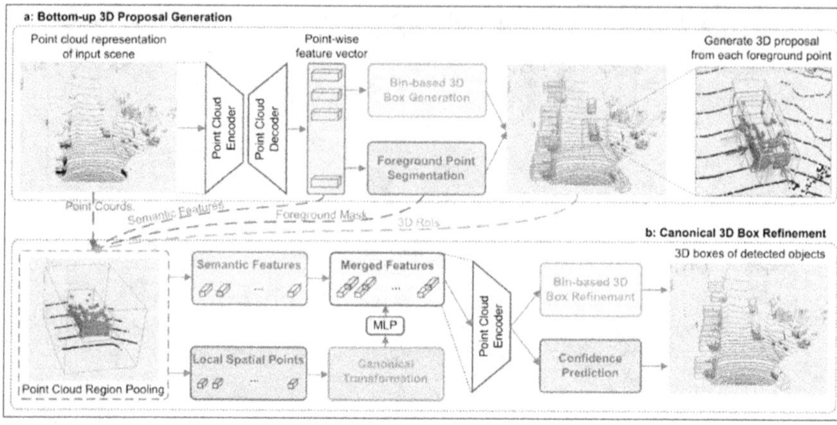

Figure 17. [18] The PointRCNN architecture for 3D object detection from point clouds. The whole network consists of two parts: (a) for generating 3D proposals from raw point cloud in a bottom-up manner, (b) for refining the 3D proposals in canonical coordinates.

Bottom-up 3D proposal generation via point cloud segmentation: In order to learn pointwise features for describing the raw point cloud, PointRCNN [18] uses as a backbone network PointNet++ [16]. The network uses foreground points to gain some knowledge of the locations and orientations of the associated objects.

- Foreground points: All 3D objects' segmentation masks could be directly obtained by their 3D bounding box annotations. 3D points inside 3D boxes are considered as foreground points.

Given the pointwise features encoded by PointNet++ [16] one segmentation head is appended, for estimating the foreground mask and one box regression head for generating 3D proposals. For point segmentation, the ground-truth segmentation mask is naturally provided by the 3D ground-truth boxes. Thus, the focal loss [19] is used to handle the class imbalance problem as:

$$L_{focal}(p_t) = -a_t(1 - p_t)^{\gamma} \, log(p_t),$$

where $p_t = p$ for foreground point and $p_t = 1 - p$ otherwise.

Simultaneously with the foreground point segmentation problem, a box regression problem is handled. A 3D bounding box is represented as $(x, y, z, h, w, l, \theta)$ in the LiDAR coordinate system, where (x, y, z) is the object centre location, (h, w, l) is the object size, and θ is the object orientation from the bird's view. For the estimation of the centre location, the neighbourhood for each foreground point is being split in discrete bins along the X and Z axes instead of handling a direct regression problem.

The localization loss for the X or Z axis consists of two terms, one term for bin classification along each X and Z axis, and the other term for residual regression within the classified bin. The centre location y along the vertical Y axis, using the $L1$ loss is enough for obtaining accurate y values because y values are within a very small range.

The localization targets are formulated as:

$$bin_x^{(p)} = \left\lfloor \frac{x^p - x^{(p)} + S}{\delta} \right\rfloor, \quad bin_z^{(p)} = \left\lfloor \frac{z^p - z^{(p)} + S}{\delta} \right\rfloor,$$

$$res_u^{(p)} = \frac{1}{C}\left(u^p - u^{(p)} + S - \left(bin_u^{(p)} \cdot \delta + \frac{\delta}{2}\right)\right), \quad u \in \{x, z\}$$

$$res_y^{(p)} = y^p - y^{(p)}$$

- $(x^{(p)}, y^{(p)}, z^{(p)})$ are the coordinates of a foreground point of interest,
- (x^p, y^p, z^p) is the centre coordinates of its corresponding object,
- $bin_x^{(p)}$ and $bin_z^{(p)}$ are the ground-truth residual for further location refinement within the assigned X and Z axis,
- $res_x^{(p)}$ and $res_z^{(p)}$ are the ground-truth residual for further location refinement within the assigned bin,
- C is the bin length for normalization.

The estimation of the orientation θ and size (h, w, l) is done based on [20]. The overall 3D regression loss L_{reg} with different loss terms for training could then be formulated as:

$$L_{bin}^{(p)} = \sum_{u \in \{x,z,\theta\}} (F_{cls}(\widehat{bin}_u^{(p)}, bin_u^{(p)}) + F_{reg}(\widehat{res}_u^{(p)}, res_u^{(p)})),$$

$$L_{res}^{(p)} = \sum_{v \in \{y,h,w,l\}} (F_{reg}(\widehat{res}_v^{(p)}, res_v^{(p)}),$$

$$L_{res}^{(p)} = \frac{1}{N_{pos}} \sum_{p \in pos} (L_{bin}^{(p)} + L_{res}^{(p)})$$

- N_{pos} is the number of foreground points,
- $\widehat{bin}_u^{(p)}$ and $\widehat{res}_u^{(p)}$ are the predicted bin assignments and residuals of the foreground point p,
- $bin_u^{(p)}$ and $res_u^{(p)}$ are the ground-truth targets calculated as above,
- F_{cls} denotes the cross-entropy classification loss,
- F_{reg} denotes the smooth L1 loss.

- **Point cloud region pooling**

After having the region proposal for the 3D bounding box for refining the specific location PointRCNN [18] proposes to pool 3D points. So, each 3D box proposal $b_i = (x_i, y_i, z_i, h_i, w_i, l_i, \theta_i)$ is enlarged by a constant n and each point that is inside the enlarged 3D bounding box is being kept, alongside its features for refining the box bi. The features associated with the inside point p include its 3D point coordinates $(x^{(p)}, y^{(p)}, z^{(p)}) \in R$, its laser reflection intensity $r^{(p)} \in R$, its predicted segmentation mask $m^{(p)} \in \{0, 1\}$ from stage-1, and the C-dimensional learned point feature representation $f^{(p)} \in R^C$ from stage-1.

- **Canonical 3D bounding box refinement**

The pooled points and their associated features for each proposal are given as input to the stage-2 sub-network. Each pooled point is being transformed into the canonical coordinate system of the corresponding 3D proposal. This means that the origin is located at the centre of the box proposal and that the local X' and Z' axes are approximately parallel to the ground plane with X' pointing towards the head direction of the proposal and the Z' axis is perpendicular to X'. Y' axis remains the same as that of the LiDAR coordinate system.

The canonical transformation enables robust local spatial features learning although because it loses depth information of each object the distance to the sensor is included in the features of point p. The local spatial features are first concatenated and fed to several fully connected layers to encode their local features to the same dimension of the global features $f^{(p)}$. Then the local features and global features are concatenated and fed into a network and a discriminative feature vector is obtained for the following confidence classification and box refinement. For box proposal refinement the bin-based regression losses for proposal refinement is adopted as in stage one.

- **Fusion Scheme**

From LiDAR coordinate system to camera coordinate system: As presented in [21], for the sensors to be synchronized the timestamps coming from the LiDAR are used as a reference and each spin is considered as a frame. The LiDAR keeps rotating to collect the data and the camera is triggered every time it faces forward. To project a 3D point $X = (x, y, z, 1)^T$ in the rectified (rotated) camera coordinates to a point $Y = (u, v, 1)^T$:

$$Y = P_{rect} X,$$

$$P_{rect} = (f_u \; 0 \; c_u \; 0 \; f_v \; c_v \; 0 \; 0 \; 1 - f_u b_x \; 0 \; 0)$$

b_x denotes the baseline (in meters) with respect to the reference camera.

Note that in order to project a 3D point x in reference camera coordinates to a point Y on the image plane, the rectifying rotation matrix of the reference camera R_{rect} must be considered as well:

$$Y = P_{rect}R_{rect}X$$

R_{rect} has been expanded into a 4×4 matrix by appending a fourth zero-row and column and setting $R_{rect}(4, 4) = 1$.

Velodyne laser scanner with respect to the reference camera coordinate system is registered using [22]. The rigid body transformation from Velodyne coordinates to camera coordinates are given from:

- $R^{cam}_{velo} \in R^{3x3}$ ...rotation matrix: Velodyne → camera
- $t^{cam}_{velo} \in R^{1x3}$ ...translation vector: Velodyne → camera
- using $T^{cam}_{velo} = (R^{cam}_{velo}\ t^{cam}_{velo}\ 0\ 0)$

Hence, a 3D point x in Velodyne coordinates gets projected to a point Y in the camera image as below:

$$Y = P_{rect}R_{rect}\ T^{cam}_{velo}X$$

Fusion Scheme Architecture: So, after we perform semantic segmentation to the attacked image, the output will resemble Figure 18 image (b). Most of the vehicles are hidden from the prescription engine due to the attack on the camera sensor. In Figure 18 image (a) we can see the output of the segmentation model in the non-attacked image.

In a parallel module, the object detection model is running, which has been trained to identify only the moving objects which are mainly vehicles, pedestrians and cyclists. In Figure 19(a) we can see the 3D bounding boxes of the vehicles that have been detected on LiDAR sensor data. Most of the vehicles have been identified from the model. After obtaining the coordinates of the 3D bounding boxes, they are being projected to the images plane in order to correlate the 2D image segmentation output and the 3D object detection output (Figure 19(b)).

More specifically in order to correlate the two outputs, we isolate the region of the projected 3D bounding box to the image. We consider that the isolated region belongs to a specific class (vehicle, pedestrian, cyclist). We isolate respectively the same region from the segmentation mask. Finally, we compare the two outputs in order to estimate the overlap between the two segmentation masks. If the object has been detected by both of the modalities the overlap should be high enough. Structural similarity index measure (SSIM) [23] is used for the comparison and should be above 0.6. The scheme for comparing the two outputs is presented in the Figure 20.

Figure 18. Image (a) is the output of the segmentation model to the non-attacked image while in the image (b) is the output of the segmentation model to the attacked image. The red mask indicates the detected vehicles.

Figure 19. (a) The output of PointRCNN to 3D space in the LiDAR coordinate system. (b) The projected output of PointRCNN to the image plane.

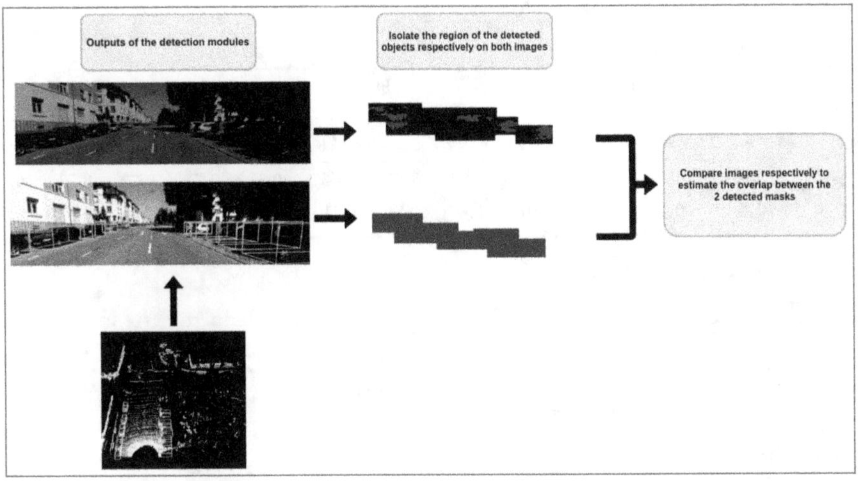

Figure 20. Comparison scheme. The 3D isolated projected objects are compared respectively with the segmentation output.

The proposed pipeline aiming to provide improved situational awareness to the user consists of two modules. The first one refers to the robust image segmentation model and the second to the 3D object detector using the LiDAR sensor. Hence, taking into consideration the first model, some evaluation metrics are shown below. Multiple attacks were implemented with different sizes of perturbation for evaluation purposes. Some of the implemented attacks on the camera sensor are shown below:

- Untargeted attacks:
  - BIM
  - FGSM
  - PGD
- Targeted attacks referring to cars:
  - LinfPGD
  - MomentumIterative

The Intersection over Union (IoU) is the primary metric used in evaluating segmentation outputs. The evaluation of the robust segmentation model of DeeplabV3 on the KITTI benchmark is illustrated in Tables 4 and 5. Table 4 refers to untargeted attacks and Table 5 to targeted attacks.

From Tables 4 and 5, we observe that our goal to provide a robust segmentation model for adversarial attacks in 2D space has been achieved. Next, some evaluation results of the 3D object detector coming from the LiDAR data are illustrated.

The evaluation of the 3D object detection model PointRCNN tested on the KITTI benchmark is shown in Table 6. For the 3D detection of car and cyclist,

Table 4. Evaluation results (IoU) for untargeted attacks between robust and original segmentation.

Resnet = 152	**BIM**				**FGSM**				**PGD**			
Model	e = 2	e = 4	e = 8	e = 16	e = 2	e = 4	e = 8	e = 16	e = 2	e = 4	e = 8	e = 16
Robust	**0.42**	**0.1**	**0.03**	**0.03**	**0.75**	**0.71**	**0.66**	**0.5**	**0.74**	**0.72**	**0.72**	**0.52**
Original	0.13	0.02	0.01	0.01	0.67	0.51	0.21	0.03	0.69	0.53	0.17	0.03

Table 5. Evaluation results (IoU) for targeted attacks between robust and original segmentation model.

	LinfPGD							
Resnet = 152	**target = 2**				**target = 7**			
Model	e = 2	e = 4	e = 8	e = 16	e = 2	e = 4	e = 8	e = 16
Robust	**0.75**	**0.7**	**0.62**	**0.5**	**0.75**	**0.71**	**0.63**	**0.46**
Original	0.69	0.58	0.37	0.17	0.69	0.53	0.26	0.1
	MomentumIterative							
Resnet = 152	**target = 2**				**target = 7**			
Model	e = 2	e = 4	e = 8	e = 16	e = 2	e = 4	e = 8	e = 16
Robust	**0.71**	**0.68**	**0.62**	**0.35**	**0.72**	**0.68**	**0.61**	**0.34**
Original	0.65	0.5	0.17	0.02	0.65	0.37	0.12	0.02

PointRCNN method outperforms previous state-of-the-art methods with remarkable margins on all three difficulties and ranks first on the KITTI test board among all published works at the time of its submission, although most of the previous methods use both RGB image and point cloud as input. For pedestrian detection, compared with previous LiDAR-only methods, our PointRCNN method achieves better or comparable results, but it performs slightly worse than the methods with multiple sensors.

Finally, some more experiments on the KITTI dataset were conducted to evaluate the overall proposed pipeline, that combine the 2D image segmentation with the 3D object detector. Hence, the LiDAR module is triggered only when an external attack to the camera sensor has been detected. In a safe situation, the final result is coming only from the camera sensor, as we can observe from the green cells from Table 7. On the other hand, when a dangerous situation has been detected, the output of the 3D detector is given to the output of the perception engine, ignoring

Table 6. [18] Performance comparison of 3D object detection with previous methods on KITTI test split by submitting to the official test server. The evaluation metric is Average Precision (AP) with IoU threshold 0.7 for car and 0.5 for pedestrian/cyclist.

		Car (IoU = 0.7)			**Pedestrian (IoU = 0.5)**			**Cyclist (IoU = 0.5)**		
Method	**Modality**	**Easy**	**Moderate**	**Hard**	**Easy**	**Moderate**	**Hard**	**Easy**	**Moderate**	**Hard**
MV3D	RGB+LiDAR	71.09	62.35	55.12	–	–	–	–	–	–
UberATG-ContFuse	RGB+LiDAR	82.54	66.22	64.04	–	–	–	–	–	–
AVOD-FPN	RGB+LiDAR	81.94	71.88	66.38	50.80	42.81	40.88	64.00	52.18	46.61
F-PointNet	RGB+LiDAR	81.20	70.39	62.19	51.21	44.89	40.23	71.96	56.77	50.39
VoxelNet	LiDAR	77.47	65.11	57.73	39.48	33.69	31.51	61.22	48.36	44.37
SECOND	LiDAR	83.13	73.66	66.20	51.07	42.56	37.29	70.51	53.85	46.90
PointRCNN	**LiDAR**	**85.94**	**75.76**	**68.32**	**49.43**	**41.78**	**38.63**	**73.93**	**59.60**	**53.59**

Table 7. Evaluation results (IoU) fusing multiple sensor data.

Data Type		**IoU Camera**		**IoU Camera + LiDAR**		**Samples**
Normal		**76**		**76**		250
Attacked			**eps**			
		2	16	2	16	
Untargeted Attacks	FGSM	**75**	**50**	**75**	**82**	250
	PGD	**74**	**52**	**74**	**81**	
	BIM	**42**	**3**	**82**	**81**	
Targeted Attacks	Momentum Iterative	**71**	**35**	**71**	**80**	
Fused Model Accuracy				**78**		500

the camera results. In the latter cases, the model performs better as we can observe from orange cells in Table 7.

Conclusions and Future Work

Throughout CARAMEL's development and experimentation phase, it was verified that the cyber attack detection and mitigation engine, developed during the project, can provide system immunization at a level of high caliber. The consortium's solution in pillar 1 was mainly geared towards sensor-oriented approaches for mitigating the attack.

More specifically, the detection and mitigation of the cyber-attacks were quite efficient in restoring the attack in numerous Autonomous Driving Functions including fully automated Parking. Regarding the in-vehicle location spoofing attack detection solution, it is important to understand and consider the anti-spoofing mechanisms already present in commercial GNSS receivers that are integrated into autonomous vehicles (e.g., timestamp checks to detect time synchronization attacks). The CARAMEL's solution was enhanced to go beyond existing schemes of detecting and mitigating fine-grain attacks that introduce small bias in the GPS locations to avoid detection. Finally, with regards to cyber-attacks on camera/lidar sensor, the consortium has developed efficient schemes of attack mitigation based on Dense and Sparse Prior models using data from a single sensor as well as utilizing the complementarity of additional sensors. That latter approach has proven to contribute significantly in making CARAMEL's solution embedded friendly. However, there still needs more work to be done on supporting multiple deep learning models and facilitating their parallel execution in real-time. According to the experience gained, as a further step for robustifying cyberattack mitigation efficiency, the community should provide solutions, where data exchanged by neighboring traffic agents are being used to mitigate the attack. In this way, scene understanding measurements contributed by the neighboring traffic agents can be used as priors in the process of reversing the attack. Thus, embodying V2X communications in the cyberattack evaluation and mitigation phase.

Due to the very strong demand for safety solutions by the equipment manufacturers, we expect a significant increase in the business in this area. Innovative solutions and competence demonstrations are added on top. Since we are product-neutral, we expect the proliferation of CARAMEL technologies and thus further indirect business.

References

[11] Jason Ku, Melissa Mozifian, Jungwook Lee, Ali Harakeh, and Steven Lake Waslander. Joint 3d proposal generation and object detection from view aggregation. CoRR, abs/1712.02294, 2017.

[12] Xiaozhi Chen, Huimin Ma, Ji Wan, Bo Li, and Tian Xia. Multi-view 3d object detection network for autonomous driving. In Proceedings of the IEEE Conference on Computer Vision and Pattern Recognition, pages 1907–1915, 2017.

[13] Bin Yang, Wenjie Luo, and Raquel Urtasun. Pixor: Real-time 3D object detection from point clouds. In Proceedings of the IEEE Conference on Computer Vision and Pattern Recognition, pages 7652–7660, 2018.

[14] Shuran Song and Jianxiong Xiao. Deep sliding shapes for amodal 3D object detection in rgb-d images. In Proceedings of the IEEE Conference on Computer Vision and Pattern Recognition, pages 808–816, 2016.

[15] Yin Zhou and Oncel Tuzel. Voxelnet: End-to-end learning for point cloud based 3d object detection. CoRR, abs/1711.06396, 2017.

[16] Charles Rui Zhong Tai Qi, Li Yi, Hao Su, and Leonidas J Guibas. Pointnet++: Deep hierarchical feature learning on point sets in a metric space. In Advances in Neural Information Processing Systems, pages 5099–5108, 2017.

[17] Charles R Qi, Hao Su, Kaichun Mo, and Leonidas J Guibas. Pointnet: Deep learning on point sets for 3D classification and segmentation. Proc. Computer Vision and Pattern Recognition (CVPR), IEEE, 1(2):4, 2017.

[18] S. Shi, X. Wang and H. Li, "PointRCNN: 3D Object Proposal Generation and Detection From Point Cloud," 2019 IEEE/CVF Conference on Computer Vision and Pattern Recognition (CVPR), Long Beach, CA, USA, 2019, pp. 770–779, doi: 10.1109/CVPR.2019.00086.

DOI: 10.1561/9781638280613.ch2

Section 2

V2X Connected Mobility

Introduction

The so called V2X (Vehicle-to-Everything) communications enable vehicles to communicate with the road infrastructure and other road users (vehicles, scooters, bikes or pedestrians) and to have a more accurate knowledge of their surrounding environment that can improve the traffic safety and provide new Intelligent Transportation Systems (ITS). European Telecommunications Standards Institute (ETSI) ITS-G5 suite in Europe and the Wireless Access in Vehicular Environments (WAVE) in the US define all standards and protocols to provide numerous ITSs.

This relatively new V2X landscape that enables novel applications based on the exchange of instantaneous position/trajectories/events between surrounding cars, can be exploited for cyberattacks. So, the authenticity, confidentiality, and integrity of the V2X messages need to be ensured. CARAMEL proposes to achieve this objective by developing a secure multi-technology On Board Unit (OBU), which is the telecommunications unit embedded in each vehicle. This OBU contains a tamper-proof hardware security element to store key material, thus preventing any illicit access to this material. The OBU module also integrates a secure ETSI-G5 stack that allows to generate and validate properly signed V2X messages. Additionally, in order to provide the digital certificates to sign these messages, CARAMEL has developed a complete Public Key Infrastructure (PKI) system following the latest ETSI security standards for ITS. The major innovation in this aspect is the possibility to revoke certificates of vehicles under attack and distribute Certificate Revocation Lists (CRL) among the remaining vehicles of the system.

On the other hand, V2X connectivity is currently provided by different technologies (IEEE 802.11p, LTE-PC5, NR-V2X, standard cellular networks), which introduce the problem of incompatibility between vehicles using different radio standards. CARAMEL proposes an interoperability solution based on Multi-Access Edge Computing (MEC) servers placed at the roadside with radio infrastructure serving two different radio technologies. The previously described OBU features two communication technologies, namely the IEEE 802.11p and the LTE-Uu modules.

V2X networks, apart from data packet exchange, offer localisation service to autonomous and connected vehicles. There are some kinds of attacks that target directly the localisation services; therefore, countermeasures have to be provisioned. Today, location information is readily available in a high number of vehicles through Global Navigation Satellite Systems (GNSS), which monitor and process the satellite signals on their onboard hardware receivers. Relying on over-the-air signals for determining the vehicle's location makes GNSS-based localisation services for autonomous/connected vehicles susceptible to well-known adversary attacks that have been demonstrated successfully on mobile devices and Unmanned Aerial Vehicles (UAV). These attacks include jamming and location spoofing that both have high feasibility, do not require physical access to the vehicle, cannot be easily detected by the driver, have high probability of success, and both create hazards that may have a catastrophic impact to the vehicle itself and the surrounding traffic network. The objective of CARAMEL is also to study and analyse the launching and evolution of these attacks against autonomous/connected vehicles, by developing efficient and reliable detection mechanisms, and to propose effective countermeasures.

Threats/Problems Considered/Detected

CARAMEL addresses the functional, security and privacy issues of the V2X Communications to provide a secure environment for ITS applications. In particular, the objectives achieved by CARAMEL are:

1. The first requirement is to provide the necessary infrastructure for interoperability of radio communications bearing in mind that there are different candidates for the radio technology. The first to appear was supported by the

IEEE and is named 802.11p, its evolution is the non yet standardized IEEE 802.11bd. Then, there are the standards supported by the 3GPP, they are the LTE-PC5 and the newer NR-V2X. All these standards enable direct Vehicle-to-Vehicle (V2V) communications. Another option is to rely on standard cellular communications using a Vehicle-to-Infrastructure-to-Vehicle (V2I2V) approach, in this case, we can use LTE-Uu interfaces or more advanced versions such as 5G.

2. The second addressed issue is the provision of a complete system that enables to verify the authenticity of the transmitted messages through a Public Key Infrastructure (PKI) that distributes and revokes certificates.
3. The third objective has been the development of an OBU with an anti-tamper Hardware Security Module (HSM) to store sensitive data required to secure the V2X communications, such as Enrolment Certificates and credentials (private keys) from Authorisation Tickets (AT).
4. Next, as most ITS messages rely on the geographic position of the vehicles, CARAMEL has to ensure that vehicle's position information is trustful and reliable. For this reason, a location spoofing attack detection system has been developed.
5. Finally, the fifth objective is to provide a system that enforces vehicle privacy through appropriately choosing the instants in which vehicles should change the certificate used to sign their packets to prevent being tracked.

Solution Design: V2X Technologies and Interoperability

General Architecture

The general architecture of the V2X system is divided into 4 blocks, as shown in Figure 1. This section is mainly focused on the description of the Roadside Infrastructure and its elements, and the OBU inside the cooperative car. Some of the other blocks appearing in this image are addressed in other sections.

Elements of the remote infrastructure and the attacker are also considered in the general picture, for the implications they may have in the secure identification processes featured during V2X communication and specifically, during an attack.

Cooperative Car: OBU and Antihacking Device

An On-Board Unit (OBU) is the telecommunication unit embedded in the standard cooperative vehicles and provides secure communication functionalities. One of the goals of CARAMEL is to develop a completely functional and secure OBU that provides the hardware for the secure V2X communications.

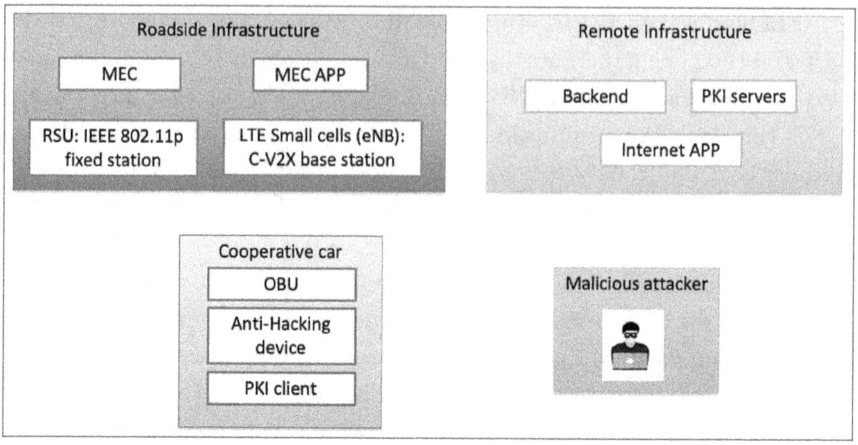

Figure 1. Global components of V2X system.

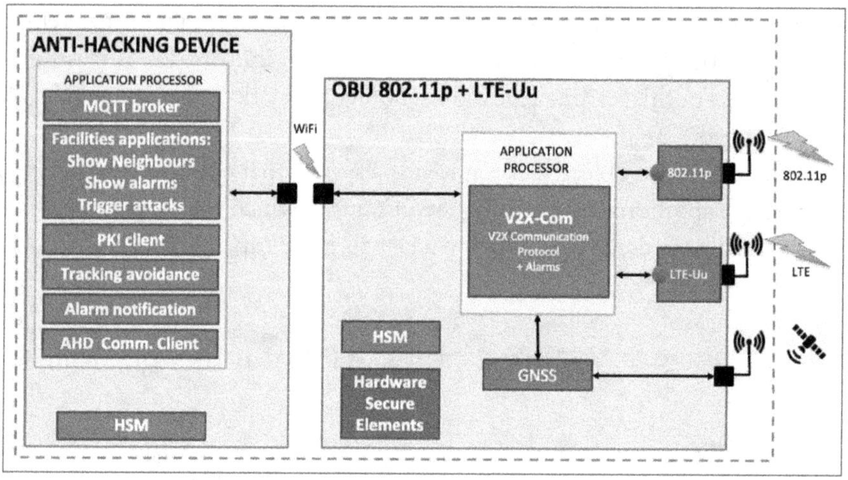

Figure 2. OBU and anti-hacking device components.

The OBU architecture is shown in Figure 2. The OBU has an application processor that hosts the V2X communication protocol stack software and a module to detect tampering attacks in the OBU's hardware. This application processor is connected to a Hardware Security Module (HSM) that is in charge of storing all communication sensitive information. The processor is also connected to a GNSS receiver that provides positioning information of the OBU to the system. Finally, the designed OBU is able to use two different radio communication technologies: IEEE802.11p, which is used for V2X communication following the standards of the ETSI G5-ITS protocols, and LTE-Uu, which is used to provide IP connection to the MEC, to the PKI servers and to the backend. In real life, all connected vehicles support LTE-Uu technology, but only some of them support

IEEE802.11p, so this combined hardware is meant to represent both types of vehicles.

The OBU security features are enhanced by the so-called "Anti-Hacking Device" that is in charge to run applications that detect malicious attacks and functional misbehaviour using pre-trained Machine Learning (ML) models and execute applications that may not fit in the OBU due to its small memory capacity. In our final prototype, it runs (i) an MQTT broker to enable communication between internal processes, (ii) applications on the Facilities layer as to show the current neighbour vehicles and alarms to the driver, and enables to trigger attacks for demonstration pourposes, (iii) the PKI client, (iv) the tracking avoidance software, (v) the software in charge to notify alarms to the backend, and (vi) a software to manage the anti-hacking HSM to secure communications between the anti-hacking device module and the OBU module.

A more detailed description to introduce each one of the elements of the OBU is given next.

Hardware Security Module (HSM)

One of the possible attack vectors to V2X infrastructure is to steal sensitive data or cryptographic keys from a vehicle's OBU. To counter this attack, trustworthy, unforgeable and non-copyable identities must be established. This is achieved by integrating an HSM into the OBU that serves as a repository for private key data (for authentication and encryption purposes), as well as a cryptographic processor for sensitive operations. The HSM is the main component to support the detection and avoidance of manipulation of the integrity of the device by providing or enabling functions such as tamper detection.

V2X Communication Protocol Architecture (V2X-Com)

This element contains the software package that enables the OBU to generate Facilities layer messages encapsulated on the Basic Transport Protocol (BTP) and the GeoNetworking protocol (GN). CARAMEL will use the open-source framework Vanetza that will be properly extended to perform all security and privacy related functionalities.

ITS Applications

This element represents any Intelligent Transportation Systems (ITS) application running on the vehicle. The CARAMEL has developed applications for sending and receiving Cooperative Awareness Messages (CAM). In a future precommercial stage, the addition of other V2X messages as Decentralized Event Notification Messages (DENM), Signal Phase And Timing Extended Message (SPATEM), MAP Extended Message (MAPEM), etc. can be easily performed.

PKI Client

The PKI function contains all software functions to interact with the PKI servers and manage the registration and authorization procedures, as well as to obtain the pseudonymous Authorisation Tickets (ATs) and store them into the HSM according to ETSI standards. It is also in charge of receiving the certificate revocation lists (CRL).

Tracking avoidance

The use of ATs to sign messages introduces an attack vector to the privacy of vehicles. Using the same AT for a long time, it enables to track the way performed by a vehicle. To minimize this effect, there is the possibility to change the used AT from time to time. This software module computes at which time it is better to change the AT to make it more difficult for eavesdroppers to track a vehicle.

Alarm notification

This element collects different alarms detected in the OBU or in the anti-hacking device and notifies the the MEC and the backend, who in turn can evaluate the risk and take decisions as, for example, to revoke the vehicle's certificates.

Radio interfaces (IEEE 802.11p and LTE-Uu)

Radio interfaces are used in CARAMEL for three purposes:

(1) For connecting OBUs to the PKI servers to obtain the ATs before being able to transmit ITS messages and for real-time management of certificates. For this purpose, LTE-Uu will be used.
(2) To notify the MEC when the anti-hacking device or the OBU's HSM detect that the vehicle is under attack. For this purpose, LTE-Uu will also be used.
(3) For data transmission between vehicles. To reduce latency during ITS message transmission, these communications will preferably use direct V2V connections through 802.11p interface. Nevertheless, in the first stages of ITS adoption, not all vehicles will be equipped with this technology. Some cars will only have the LTE-Uu interface. These cases will use the forwarding assistance of the MEC, which consists in forwarding messages between both technologies (802.11p and LTE-Uu), and also to provide broadcast addressing for LTE-Uu using its inherent unicast behaviour.

CARAMEL project has developed a complete OBU device which is able to operate with both technologies, and, additionally, in order to simulate LTE-only OBUs, some of them have the 802.11p Network Interface Card (NIC) deactivated for testbed purposes.

Network capabilities: Automotive ethernet, CAN, Wifi 802.11

The OBU is equipped with several communication interfaces that enable networking capabilities to the vehicle. This is part of the In-Vehicle Network (IVN) interface. This includes 1000Base-T1 Ethernet interface, which defines Gigabit Ethernet over a single twisted pair for automotive and industrial applications, a WiFi interface, compliant with IEEE802.11a/b/g/n/ac, 5G MIMO and RSDB (Real Simultaneous Dual Band), and finally a Controller Area Network (CAN) bus interface.

Hardware Secure Elements

These elements are included to protect the OBU from tamper attacks, through box opening detection, active hardware protection of susceptible signals, and environmental sensors to prevent fault injection attacks. When the OBU detects an attack, there is a tamper response and the system is enabled to protect sensible data. Logical methods are also used to prevent firmware manipulation. In order to comply with these security functional requirements, several tamper protection layers have been applied on the different OBU interfaces based on hardware actuations.

Roadside Infrastructure

The roadside infrastructure is composed of the IEEE 802.11p Road Side Units (RSU), the small cells pertaining to the LTE network, and the backhaul network to connect these devices to the MEC (Figure 3). The connection links between RSUs and small cells to the MEC can be performed by different means, wired or wireless. In the proposed testbed, these links will be implemented using wired Ethernet and through a VLAN capable switch.

RSUs

The RSU is the element that receives IEEE 802.11p messages transmitted by vehicles and forward them to the MEC, and transmits, via its IEEE 802.11p radio transmitter, messages generated in the MEC to the vehicles with an IEEE 802.11p interface.

The RSU architecture, shown in Figure 4 includes the following elements:

OS

The Operating System (OS) running in the device is Linux.

Management

This element enables the remote connection and administration of the device. The operations that need to be performed are basically to activate and deactivate the IEEE 802.11p transceiver and basic management of the network interfaces. For these basic functions a client-server ssh (Secure Shell) is used.

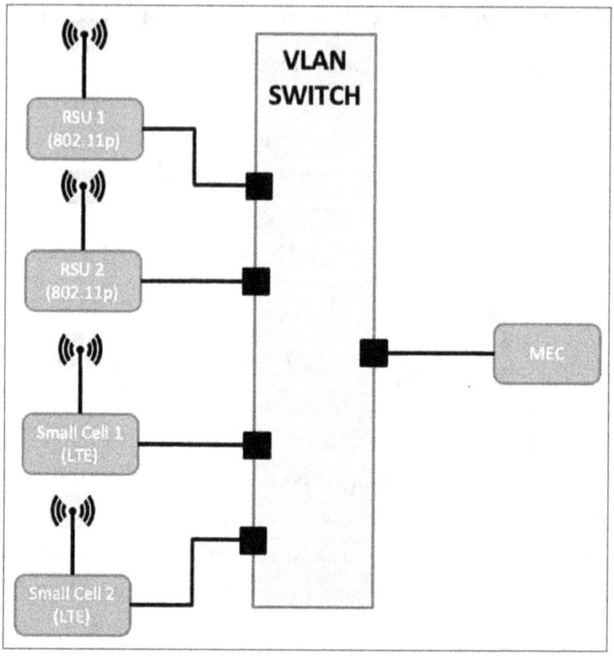

Figure 3. Connection of radiating infrastructure to the MEC.

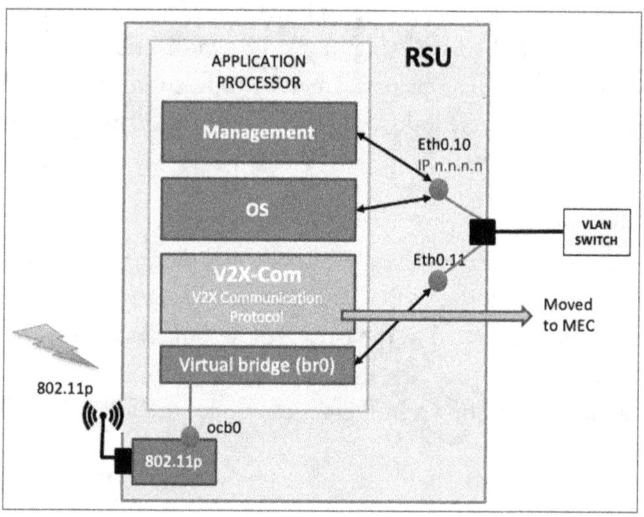

Figure 4. IEEE 802.11p RSU components.

Ethernet connection

The RSU is connected to the MEC using a wired Ethernet through a VLAN capable Ethernet switch. The RSU needs two virtual network interfaces associated to the same physical interface: one standard IP network interface to connect the previous ssh service and perform other maintenance functions as software updates, and another layer 2 network interface, without IP address, which is used to transparently forward IEEE 802.11p frames between the IEEE 802.11p transceiver and the Ethernet network which, using a VLAN configuration, transmits these frames directly to the V2X-Com module executed in the MEC.

Virtual bridge

This element performs the automatic forwarding between the IEEE 802.11p network interface and the virtual layer 2 Ethernet network interface, connected with a VLAN to the V2X-Com module executed in the MEC.

IEEE 802.11p transceiver

This element implements the IEEE 802.11p radio standard, and transmits and receives IEEE 802.11p frames to a preconfigured channel in the band ITS-G5. To implement it, we have used the Network Interface Card (NIC) WLE200NX by Compex (Figure 5). It provides dual-band MIMO technology, offering rates of up to 300 Mbps working in 802.11n and equipped with the Qualcomm Atheros AR9280 chipset. Using the Atheros XB92 as a reference model, it is capable of working in the 2.4 GHz and 5 GHz bands in 802.11a/b/g/n, offering maximum transmission power per chain of 18 dBm and 17 dBm, respectively. It has support for WPA2 encryption (IEEE802.11i) and 802.1X authentication, transmit power control (IEEE 802.11h), Dynamic Frequency Selection (DFS) technology and supports additional regulation domains (IEEE 802.11d). Regarding its interfaces, it has a miniPCI-E interface and two U.FL connectors; its energy consumption is between 1.5 W (standby) and 3.7 W (at maximum performance).

Because it is part of the Atheros 9K family, its mPCI-e connector and its operating frequency range it is chosen for the deployment of the CARAMEL RSUs to

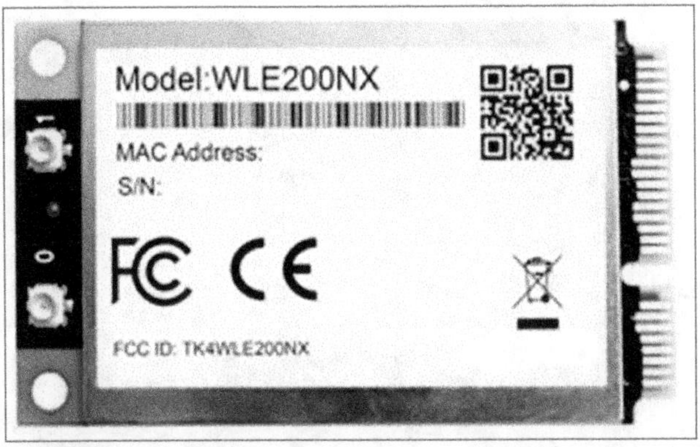

Figure 5. Compex WLE200NX IEEE 802.11n and IEEE 802.11p (ath9k).

send and receive messages through the 5.9 GHz ITS-G5 band using IEEE 802.11p. The Atheros 9k series allows to carry out the necessary changes in Kernel and device drivers so that it can work in a stable way in the 5.9 GHz band in OCB mode.

V2X-Com is implemented in the MEC

The V2X-Com module implements the networking and transport protocol layer of the ETSI GeoNetworking protocol architecture (GeoNetworking and BTP protocols), serializes and parses the V2X messages of the Facilities layer (CAM, DENM, MAPEM, SPATEM,…) and computes and checks the digital signatures using the available Authorization Tickets.

Usually, this element is implemented inside the RSU, as it happens with commercial RSUs, nevertheless, in CARAMEL, this element is moved to the MEC because it has a greater computation power and it is closer to the processes that consume these Facilities layer messages.

LTE small cells

CARAMEL project is built under the premise that vehicles have an LTE connection to access vehicular services, and some of them, those that do not have an IEEE 802.11p network interface, will also transmit their V2X messages over this LTE connection.

Therefore, V2X services operators also need to have access to the LTE network, or deploy their own LTE network on the areas where they plan to operate their services. In CARAMEL, for demonstration purposes, we develop a small LTE operator (Figure 6) that contains 2 small cells or eNBs (evolved Node B) plus an EPC (Evolved Packet Core) which, in our case, is virtualized inside the MEC (vEPC).

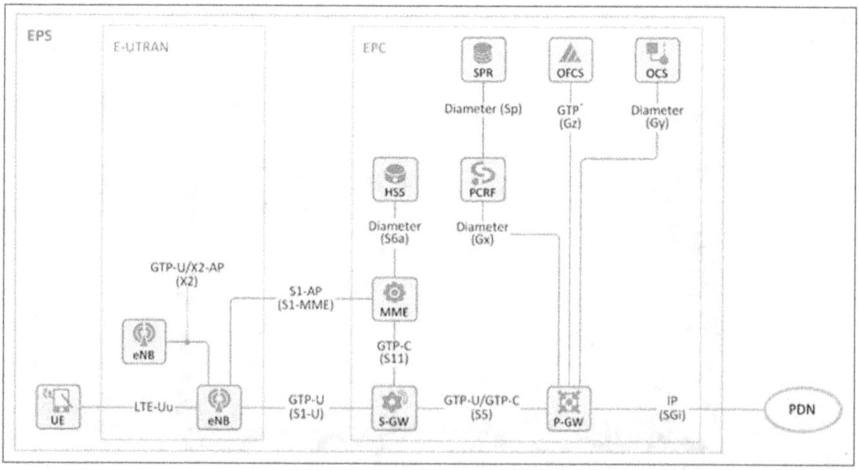

Figure 6. LTE network architecture.

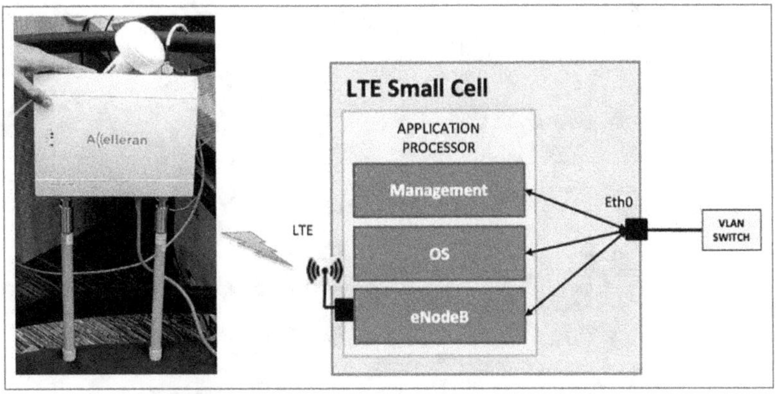

Figure 7. LTE small cell plus its components.

The two acquired small cells are from Accelleran (Figure 7), they work at the band B43 (3700 MHz – 3800 MHz) and they have the basic components: one Linux OS that can be remotely managed using standard tools as ssh, and the software to act as an eNB.

These small cells are connected using Ethernet to the MEC.

Multi-access edge computing (MEC)

Multi-access Edge Computing (MEC) is an ETSI-defined network architecture that enables cloud computing capabilities and an IT service environment at the edge of the cellular network. The basic idea behind MEC is that by running applications and performing related processing tasks closer to the cellular customer, network congestion is reduced and applications perform better. MEC technology is designed to be implemented at the cellular base stations or other edge nodes, and enables flexible and rapid deployment of new applications and services for customers. Combining elements of information technology and telecommunications networking.

In the case of CARAMEL, the MEC is the V2X interoperability node to enable communication between the vehicles and the communication base stations with the remote infrastructure comprising the PKI and the Backend. The MEC provides functions of message forwarding and distribution of revoked certificates. The MEC also has a filtering capability to filter V2X messages with revoked certificates, or obsolete in time, with data not relevant or not accurate.

The MEC architecture, shown in Figure 8, includes the following main elements:

vEPC

The virtual Evolved Packet Core (vEPC) of the LTE operator is deployed in the MEC. Among all LTE's core elements, our implementation contains the Mobility

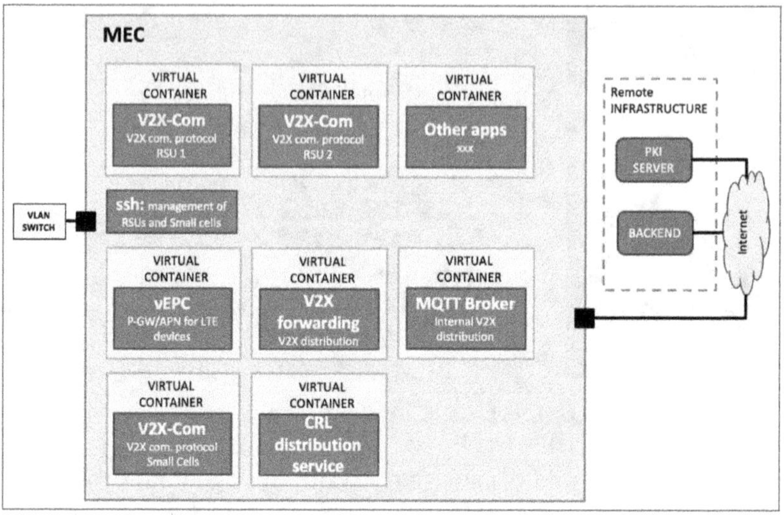

Figure 8. MEC components.

Management Entity (MME), the Serving Gateway (S-GW), the Packet Data Network Gateway (P-GW), the Home Subscription Server (HSS), and the Policy and Charging Rules Function (PCRF) which are the basic ones to enable an LTE network to operate. Finally, it provides an Access Point Name (APN) from where the OBU's applications are able to connect to Internet and to other processes running in the MEC.

V2X Communication Protocol Architecture (V2X-Com)

As it has been pointed earlier, this element, usually executed in the RSUs, has been moved to the MEC to facilitate and make it more efficient its interaction with the other modules.

The V2X-Com module implements the Networking and transport protocol layer of the ETSI GeoNetworking protocol architecture (GeoNetworking and BTP protocols), serializes and parses the V2X messages of the Facilities layer (CAM, DENM, MAPEM, SPATEM,…) and computes and checks the digital signatures using the available Authorization Tickets.

CARAMEL V2X-Com modules are based in the open-source framework Vanetza that has been properly extended to perform all security and privacy related functionalities.

There are several instances of this V2X-Com module, one for each IEEE 802.11p RSU, and another for the V2X communications through the LTE network.

MQTT Broker

The MQTT Broker is used to interchange V2X messages among applications running in the MEC. Incoming messages forwarded from the OBUs to the MEC by

RSUs and small cells, are published by the V2X-Com module to specific topics in the MQTT broker. Afterwards, they can be consumed by other applications who have subscribed themselves to these topics. Likewise, messages that the forwarding policies dictates that need to be forwarded, are published in the MQTT broker and pushed to the V2X-Com modules to be sent to the OBUs.

V2X Forwarding

This module receives all incoming V2X messages forwarded from OBUs to the MEC by RSUs and small cells, and decides to which other OBUs they need to be forwarded. Forwarding rules are based on Region of Interest (ROI) and vehicle types.

The V2X-Forwarding module has access to a MEC database named Local Dynamic Map (LDM) which contains information about the current position of all recently seen OBUs. This LDM, which is updated every time that one CAM from one OBU is received, contains additional information as if the vehicle transmits V2X messages over IEEE 802.11p or LTE, the type of vehicle, if their messages are correctly signed, and other information that can be obtained from inside CAM messages.

MEC Revocation client

The purpose of this MEC Revocation service is to facilitate the communication flow of the PKI core. It receives revocation requests, decides if an attack requires a revocation of the certificates, and sends it to the Backend for statistics purposes and to the PKI server to request the revocation of the certificates.

MEC PKI client

The purpose of this service is to provide Certificate Revocation List (CRL) dissemination to all the connected OBU's. It provides a CRL synchronization mechanism to all the connected OBUs. Periodically, it sends the current version of the CRL to a specific MQTT channel so that if any client currently has a different version of the CRL it can request it from the MEC using an exposed API.

V2XCom: Software to implement the ETSI ITS G5 communication protocol stack

The module V2X-Com takes care of handling the encoding/decoding of V2X messages and the lower layer protocols like GeoNetworking and BTP, which sometimes can be accompanied with an extra header of UDP/IP.

To do so, the module is divided into two submodules for extensibility reasons; the so-called "encoder" module; which takes care of the encoding and decoding of standardized V2X messages (like the ones standardized by SAE DSRC J2735, and the ones standardized by the ETSI like CAMs or DENMs). And the BTP-Geonet module which takes care of assembling the BTP/GeoNetworking headers given a

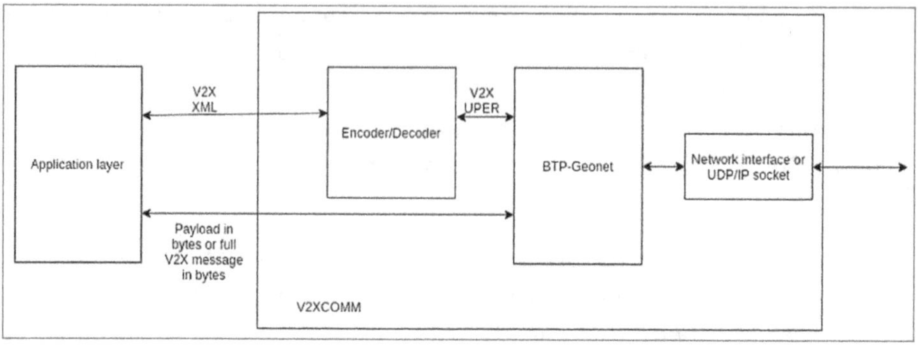

Figure 9. General architecture of an application or set of applications using the V2XCOMM module.

geodesic position coming from a GPSD, a configuration which states the destination along with any other configurable parameters that BTP and GeoNetworking headers require and the payload that has to be sent in bytes.

The communication between the different modules is done through MQTT queues handled by a broker. The broker doesn't appear in Figure 9 because it is taken for granted. Before going deeper into explaining component by component it is worthy to explain how a CAM is generated, encoded, signed and sent. The journey of a CAM starts at the application layer, where it is generated in an XML format. The XML format is friendly and readable not just by any program but also by any person. After that, the application layer sends the XML to the "Encoder" module through the corresponding MQTT queue. The "Encoder" encodes the CAM message according to the standards, so turns the XML into an encoded UPER byte array which is automatically sent to the "BTP-Geonet" module using the corresponding MQTT queue. The BTP-Geonet module, depending on the queue from which the encoded message has been received the byte array of the "encoded" CAM will construct the headers based on a previously given configuration. Also depending on this configuration the packet will be signed with the respective certificate. Once the header is generated and the packet signed, it is automatically sent to the preconfigured network interface or UDP/IP socket.

As it can be seen, the idea behind the V2XCom module is to simplify the development of V2X applications with just having to code the XML handling of the given V2X messages. On the other hand, the whole process that has been just described is completely symmetric. If a new CAM is received from the network interface, then the "BTP-Geonet" module will pre-process it, by checking the headers and dropping it if necessary. For example, it checks if the ITS-S is within the receive range of a Geobroadcasted packet or checks if the packet signature is correct. When the headers of a received packet are correct the "BTP-Geonet" module uploads the packet payload to the "Encoder" which automatically will decode the

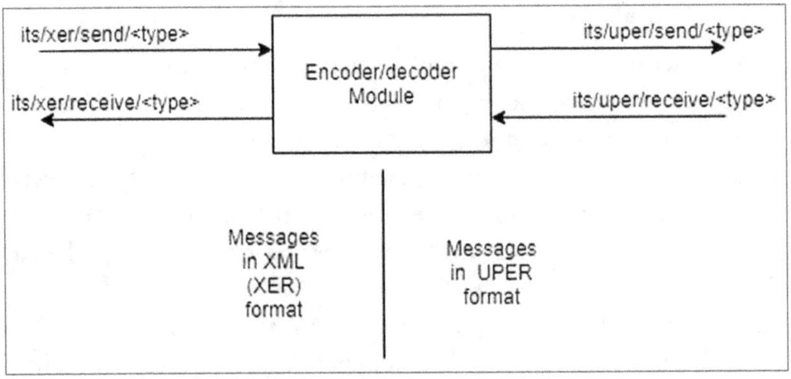

Figure 10. The Encoder/Decoder module with an example of MQTT queue's names.

UPER byte array to the corresponding XML version and also upload the decoded payload to the application layer.

As can be seen in Figure 9 there is also a fast track for non-V2X messages that don't have to be coded or decoded. So, any array of bytes can be sent with a BTP and GeoNetworking header when wanted.

Encoder/Decoder Module

Despite the function of the Encoder/Decoder module described above, al the modules of the V2X-Com component can have utility by themselves. So, it's interesting to explain deeper what the module is capable of. The module is written in C/C++ and consists of two MQTT clients; one that sends and receives XML messages, and another that sends and receives UPER packets. All of the packets go through an "encoder" class which is capable of encoding or decoding each message. To make the module completely coupled with the other modules, the names of the queues have to be previously configured.

Figure 10 describes a scenario where any XML V2X message is published at the "its/xer/send/<type>" and automatically the encoded version is published at the "its/uper/send/<type>", so, if a CAM has to be encoded it will be published at the queue "its/xer/send/cam" and automatically encoded and published at "its/uper/send/cam".

The types of messages that the module supports are customizable and if the V2X messages are standardized using ASN1 they are quite easy to add as compatible messages.

As this document is written, the compatible messages are the following ones:

- CAM
- DENM
- MAPEM
- SPATEM

BTP-Geonet Module

The BTP-Geonet module takes care of all the logic and implementation regarding the transport and network layer in a given C-ITS system. To do so, as stated above, it receives the payload of the messages to be sent, a configuration of where those messages are to be sent (for example, if they are GBC or SHB, the certificate that has to be used to sign them, or the area on which are to be sent), the current GPS position and a connection to the lower-layer network interface or the UDP/IP socket if wanted.

To accomplish all the requirements expected for the current project; the module requires dynamic and flexible configurability. Also, it is taken for granted that multiple instances of the BTP-Geonet module will be running at the same time. Not just because it will require ensemble messages considering different locations, but also because the deployment of the project will require handling different interfaces/sockets (outputs) for the crafted V2X messages. So, each instance, when launched will be configured to be hooked to a specific position (that will be prefixed or constantly scrapped from a GPSD), a network interface or UDP/IP socket (as the output for the generated messages) and the name of the instance (which will be used later at the MQTT queues naming convention).

Once the static characteristics of the configuration are settled at launch, the more dynamic ones (BTP port, GN destination…) require a different treatment. For that reason, it handles each configuration for sending V2X messages with the BTP and GeoNetworking protocol as a "socket". Which can be opened through passing the new socket configuration in JSON format through MQTT at the queue named "<instance_name>/open". Where the "<instance_name>" is the previously configured name for the instance. And as an example for opening a CAM socket the JSON would look like as follows:

```
{
        "btp" : {
                "port": 2001
        },
        "geonet":{
                "certificate": "/files/ticket.cert",
                "key": "/files/ticket.key",
                "cert_chain": ["/files/root.cert", "/files/aa.cert"],
                "mode" : "shb",
        },
        "general":{
                "timeout": 600,
                "name" : "cam"
        }
}
```

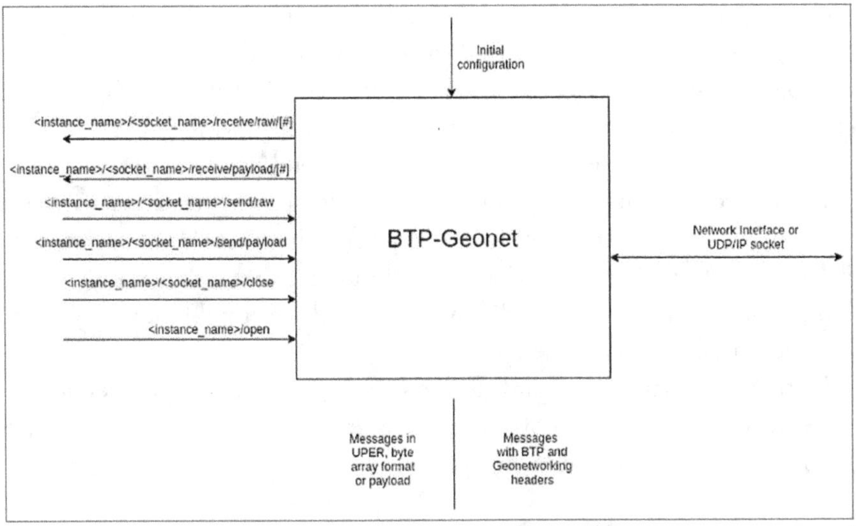

Figure 11. The BTP/Geonet module with an example of queues names.

Where the BTP port is configured with the 2001, the packet signed certificate is specified and also the name of the socket at the "general -> name" configuration parameter. So, all the communication to the recently opened socket will be done through the queues whose name starts with "<instance_name>/<socket_name>". An example of what can be done with sockets is to 'close it', and to do so, any information can be sent to the queue "<instance_name>/<socket_name>/close" and it will automatically close the socket.

Once it is known how to open/close sockets, as well as properly configure the whole module; the only thing left to see is how to send/receive messages. To send a V2X message, the payload of the message just needs to be published at the queue "<instance_name>/<socket_name>/send/payload" and the BTP-Geonet module will ensemble the headers and send the message. To forward an entire message (already signed) which doesn't need the crafting of more headers, they have to be published at the "<instance_name>/<socket_name>/send/raw" queue. Finally all V2X message received will be automatically published at the queues: "<instance_ name>/<socket_name>/receive/ raw/[#]" and "<instance_name>/<socket_name>/receive/payload/[#]". Messages published in the queue "raw" publish the entire byte array (including headers) and for the queue "payload" only the payload (This is done for forwarding reasons). And finally, the "[#]" will be a number identifier for the packet, so there is a way of relating payloads to the whole message. Figure 11 resumes the inputs/outputs of the module.

An example to show the flow of setting up the module, sending and receiving CAM message would be: A BTP-Geonet needs to be launched configured (by command line or docker) to be hooked to an 802.11p interface, to scrap the positions from a running GPSD service and to be named "instance1". Then, the socket for CAMs needs to be opened through publishing the JSON message shown above to the "instance1/open" MQTT queue (remember that the name of the new socket is "cam"). Finally, to send a CAM message only requires to publish the UPER payload of the CAM at the queue "instance1/cam/send/payload".

The architecture is completely symmetrical, so, if the socket receives a CAM message, it will publish it automatically to the queues: "instance1/cam/receive/raw/1" and "instance1/cam /receive/payload/1" (considering that's the case of the first CAM published). Finally, the socket will be closed when publishing anything to the queue "instance1/cam/close".

The current explanation regards to the first steps of the software development and despite what's described above has already been developed, the naming or some of the features are still bound to change.

Encapsulation of V2X Messages Over IP

V2X messages with GeoNetworking and BTP headers are intended to be transmitted directly over a layer 2 interface. Nevertheless, in some situations, the only chance is to transmit them over an IP interface. The V2X-Com module contemplates this possibility which is the one to be used when V2X messages are transmitted by OBUs with a single LTE-Uu interface.

The approach taken by caramel is to use an UDP socket, that is opened between the V2X-Com modules deployed in LTE-Uu only OBUs and and the V2X-Com module of the MEC that manages messages transmitted by the LTE vEPC.

Figure 12 shows the used encapsulation of such messages.

Remote Infrastructure

The remote infrastructure is formed of all servers that run in the cloud, in CAREMEL project, these are the PKI and the backend servers.

Backend

With increased use of advanced technology and external connectivity for vehicles, cyber security is becoming a major concern, and not only from financial perspective but also from customer acceptance perspective. All though there are multiple

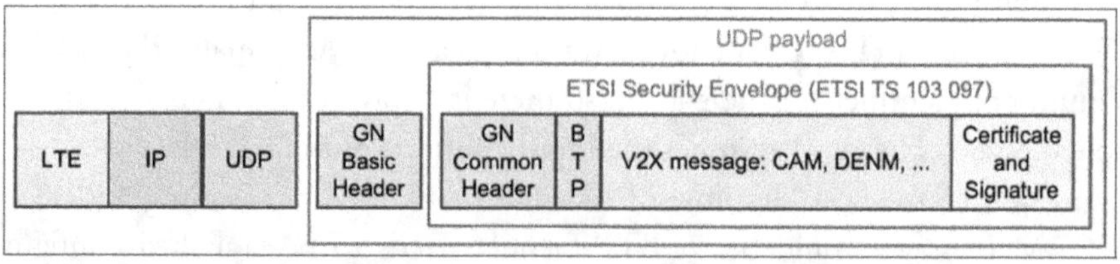

Figure 12. Encapsulation of a V2X message over IP/UDP.

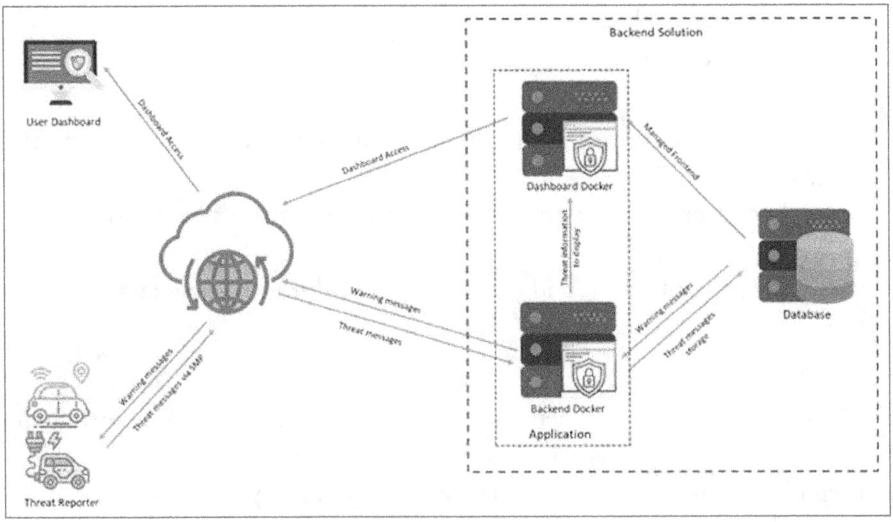

Figure 13. Backend solution system architecture.

cyber-attack prevention technologies in the market covering the known attacks, cyber-attack detection is becoming an important component for after sales of vehicles as it not only detects the known attacks but also makes it possible to recognize new attack patterns which becomes key for developing a protection mechanism.

One of the core elements of detection technology is a Backend solution for collection, detection and investigation of information related to suspicious activities in the vehicle. In CARAMEL, we have tried to cover this aspect by developing a Proof of Concept (PoC) for a Backend solution, which would be able to integrate all use cases of CARAMEL together and provide a central repository of threat information. Threats can be detected by vehicles, anti-hacking device, Intrusion Detection Systems (IDS), MEC, etc. and reported to the backend solution for investigation and information dispersion. The backend solution can be utilized by different private/public organizations to monitor, investigate and respond to the reported threats.

Figure 13 shows the system architecture for the backend solution developed as part of the CARAMEL project. It is accessible through the Internet for receiving threat messages, sending warning messages and access to the dashboard. Its features are:

- Dashboard
- Map Visualization
- Threat Visualization
- Localization of Threats on the map
    - Local Threats

    - Regional Threats
    - Pop-ups
- Tracking of New Threats
- Filtering of information is possible based on Location, Threat type and Vehicle ID's
- Sending information/warning back the vehicle operator by position or trajectory
- Clustering local threats and merging regional threats is possible

PKI servers

The PKI servers manage the identities assigned to the vehicles and the PKI architecture addresses the security requirements of confidentiality, integrity, availability, non-repudiation and anonymity, allowing vehicles to safely transmit secured messages. The PKI architecture is formed by five authorities, each satisfying a specific task, as described in Figure 14.

The Root Certification Authority (RCA) is an offline server which stores the root certificates for the entire PKI infrastructure. The RCA signs the certificate of the Online Certification Authority (OCA) and its main responsibility is to sign the different lower authorities in the PKI infrastructure. The Enrolment Authority

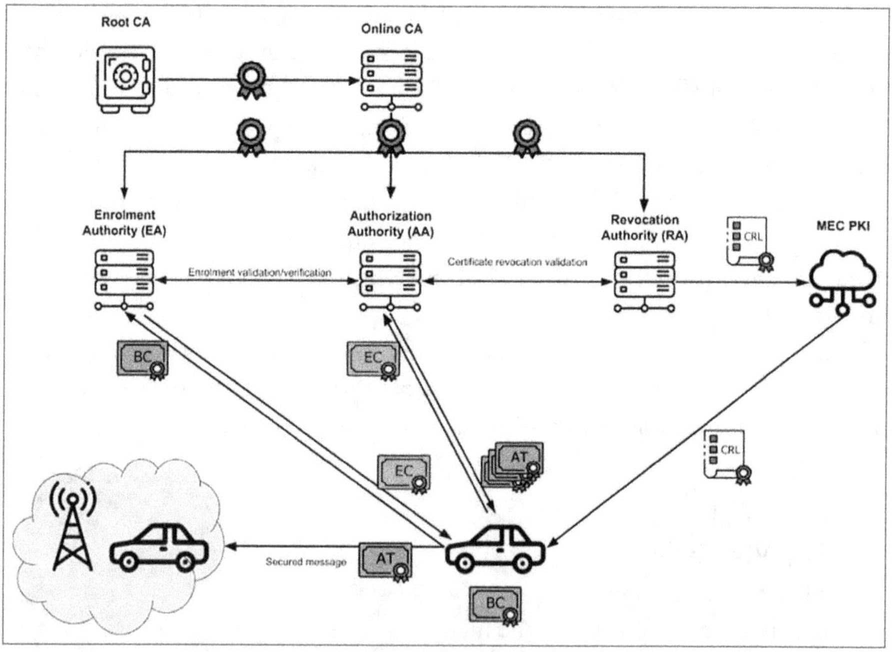

Figure 14. PKI-Toolbox high level architecture.

(EA) oversees the provisioning of necessary Enrolment Certificates (EC) to the vehicles during the enrolment phase. At the beginning of the lifetime of a vehicle, its manufacturer must request the registration of the vehicle through the Bootstrap Certificate (BC). Then the vehicle must perform an Enrolment Credential request by sending to the Enrolment Authority its BC. The EA verifies the request and in case of positive verification the EA registers the vehicle information in its database. This operation can be performed only once and only one EC for each vehicle can be issued. The ECs identify the vehicles in the long term and allow the vehicles to obtain the Authorization Tickets from the Authorization Authority. The Authorization Authority (AA) is the entity which manages the Authorization Tickets (AT). ATs are issued to ensure privacy and anonymity of the vehicles within the PKI infrastructure, identifying vehicles in the short term and allowing them to send anonymous secured messages to their surroundings. The process of revocation is managed by the Revocation Authority (RA), which processes the revocation requests from the MEC PKI and generates the Certificate Revocation List (CRL) containing the revoked ATs. The CRL then is transmitted to the MEC PKI which relays it to the vehicles, in this way instructing them on which certificates to trust. Additionally, the RA offers an internal API to check the validity of a specific AT in real-time.

The communication flows between entities and vehicles are described in Figure 14. A vehicle directly contacts the PKI Core to perform the Enrolment and Authorization process. While the Revocation process is handled by the MEC. Specifically, in the revocation process, the communication between the Core PKI and the MEC is performed using one-to-one calls using protocols such as Rest API. The revocation process includes the report of revoked vehicles by the MEC and the generation of the CRL by the Revocation Authority. Every update of the CRL is forwarded by the MEC to the vehicles using a publisher-subscriber messaging pattern such as the MQTT protocol. On the vehicle, the OBU communicates with the PKI client using direct calls to the module. In this process, the OBU interrogates the PKI client about the validity of a certificate. The PKI client checks the CRL list and provides an answer.

V2X Message Authentication and Privacy + CRL Distribution

For privacy purposes, a set of ATs, which are not linked between each other, will be assigned to the same vehicle to guarantee anonymity. The total number of short-term certificates assigned to a vehicle depends on different factors, such as the validity period of the certificates and the number of simultaneously valid certificates assigned to a vehicle to protect its privacy. These factors are analyzed in more detail in the subsequent sections, however, the total number of certificates stored on a vehicle can grow very large. The Butterfly key expansion technique grants the ability of generating batches of unlikable certificates starting from a single request, diminishing the load on the requesting vehicle. The impossibility of linking the certificates between each other guarantees anonymity but on the other hand implies that certificates must be revoked one by one. When a malicious vehicle is detected, to prevent it from communicating with its surroundings and potentially take advantage of the infrastructure, all its active short-term certificates must be revoked. This requirement contrasts the scalability requirement which demands an agile way of managing all certificates of the whole system.

Acknowledging that the anonymity requirement is abiding, and the pseudonym certificate architecture is hardly alterable, we seek to optimize the scalability requirement leaving untouched the anonymity practices. Scalability is crucial in the process of revocation of a misbehaving vehicle, as the communication of the identified misbehaving vehicle to the other vehicles in the system must be as swift as possible. The transmission of this information is performed with the broadcast of a CRL which grows in size proportionally to the number of revoked vehicles. The CRL, which is generated by the Revocation Authority and broadcasted to vehicles and RSUs, contains one entry for each certificate that is revoked. Vehicles and RSUs will use the information included in the CRL to drop all messages received that have been encrypted and/or signed with a certificate included in the CRL, without attempting to process them. Multiple certificates from a revoked vehicle can be identified as invalid, including both short and long-term certificates, and therefore included in the CRL.

Considering the number of short-term certificates assigned to a vehicle and the growing amount of revoked vehicles over time, the number of certificates to be included in the CRL can rapidly become hard to manage impacting in the bandwidth usage and processing overhead, and consequently having a negative impact on the scalability of the system and endangering its operation. For these reasons we analyzed different ways of optimizing the CRL distribution to ensure its timely delivery:

1. Binary Hash Trees: V. Kumar et al. [1] proposed a different approach to standard revocation lists, replacing them with Certificate Access Lists (CAL). Vehicles are provided with encrypted certificates, specifically each batch of certificates (certificates to be used at the same time) is encrypted with the same symmetric key.

   An entity called Certificate Access Manager (CAM) creates the encryption keys for each vehicle and each time frame. To do so, the CAM for each time frame creates a binary hash tree starting from a random seed. The depth of the tree is such that each leaf represents a vehicle. The value of each leaf, each identifying a vehicle, is the symmetric key for the decryption of the batch of certificates of the corresponding vehicle. Each valid vehicle, starting from a node of the binary hash tree, is able to compute the children nodes and reach the leaf node having as value the decryption key for the active certificates. For this reason, it is not necessary to broadcast the whole tree but only the nodes necessary to calculate the leaf nodes of the valid vehicles. Once the CAM has generated the hash tree for all the valid vehicles, it forwards the necessary nodes to obtain the decryption keys through the CAL. As a consequence, in the case where no vehicle is revoked, only the root of the binary hash tree needs to be forwarded. As the number of revoked vehicles grows, the average number of broadcasted messages is defined as: where is the number of revoked vehicles.

This approach brings the following advantages, however, this approach has some drawbacks, especially regarding the integration in the structure described by ETSI [2]. The adoption of new entities and the transition from Certificate Revocation Lists to Certificate Access Lists poses an additional challenge.

2. Binary Hash Trees + Activation Codes: M.A. Simplicio et al. [3] extended and improved BCAM [1], starting from the same main concepts. The main difference lies within the generation of pseudonym certificates. Like In the previous approach], pseudonym certificates are generated and then encrypted using a binary hash tree. However, here the authors propose the generation of codes for each vehicle and each time period using a binary hash tree. This code is called "activation code". The activation codes are integrated directly in the *butterfly key expansion* process used to create the pseudonym certificates. In both cases, the nodes of the binary hash tree which allow the calculation of the activation codes are broadcasted to the vehicles using RSUs. This new approach including activation codes brings higher performance and better security with respect to one based only on binary hash trees at the cost of additional complexity on the PKI infrastructure. An increase of performance is achieved by removing one layer of encryption and integrating the activation codes in the pseudonym certificates, saving numerous operation cycles. By adding activation codes, one drawback of the previous design is addressed by creating an extra point of collusion in the architecture. The CAM, like the Revocation Authority, learns which batch of encrypted certificates belong to the same vehicle and consequently, the CAM can collude with the Pseudonym Certificate Authority to violate those certificates' unlinkability and, hence, the users' privacy.

3. Bloom Filters: They are a commonly used probabilistic data structure that allows to efficiently compress information using the output of multiple hash functions, representing a set in a space-efficient way. Because of its intrinsic probabilistic nature, Bloom filters assessing the belonging of an item to a set could yield false positives but never false negatives. In the case of CRL distribution for VANETs, the set represented with the Bloom filter is the list of revoked certificates. Therefore, a query on the belonging of a certificate to the set will: (a) Never return that a revoked certificate is valid (false negative) and b) Occasionally detect a valid certificate as revoked (false positive). Bloom filters can be used as a drop-in replacement of standard CRLs, producing compression gains by, approximately, a factor of ten, depending on many factors. False positives, however, pose a challenge which can be addressed in two ways: backup certificates or chained bloom filters.

After considering the previously introduced possibilities, we selected to use Bloom filters to optimize the distribution of the CRL, for different reasons:

- Bloom filters are extensively studied and have been integrated in the certificate revocation process in many fields, from smart grids and the Internet Protocol to the vehicular network itself.
- Bloom filters allow for an almost seamless integration in the current infrastructure, without the need to add additional entities to the PKI infrastructure.
- The significant compression gain derived from the space efficient structure justifies the introduction of a probabilistic data structure, moreover its drawbacks have been thoroughly studied and can be addressed.

To validate our selection we developed a simulation to showcase the benefits of Bloom Filter CRL compression and, in particular, we investigated two main objectives: the optimal configuration parameters of the Bloom Filter and the compression gain of a Bloom Filter CRL with respect to a standard CRL. In the studied simulation the BF represents the list of revoked AT and it will return a false positive with a defined probability when interrogated on whether a specific AT belongs to the set. In the case of a false positive the BF will indicate that an AT is in the set and flagged as revoked even if it is not. The challenge of false positives is tackled using backup certificates on the sender side.

Firstly, we analyzed the compression gain against the false positive probability, as a low compression gain would not justify the introduction of a BF and, finally, a high FPP that could challenge the effectiveness of the filter. Figure 15 represents the output of the first simulation. Each point on the chart represents a simulation of N=1e6 items inserted in a BF created with a different false probability input rate. The compression ratio grows with the false probability rate and because of this a good balance between the two parameters must be selected. We believe that

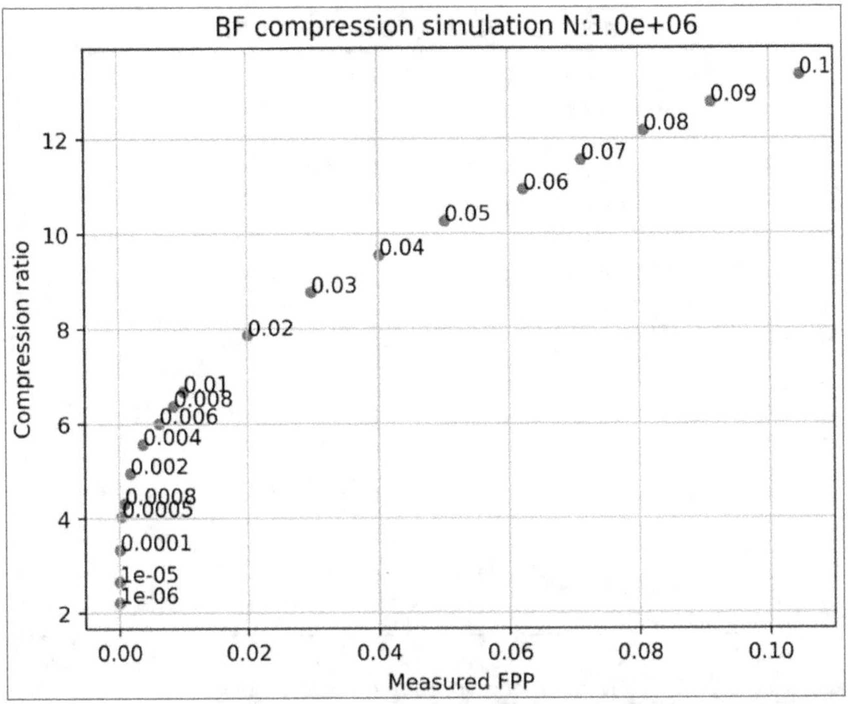

Figure 15. Compression gain against simulated (measured) false positive rate.

a FPP higher than 0.04 would be impractical in the CARAMEL context and for this reason they have not been considered in the following analysis. While a compression of a factor of 8 would mean that the BF CRL grows by one byte at every AT revoked, with and expected FPP of 0.02, this scenario would require a high number of revoked certificates, mitigating the benefits of the BF. We expect lower FPP to be beneficial to the CARAMEL scope.

Secondly, we compared in Figure 16 different BF configurations against the standard CRL. The standard CRL grows linearly with the number of revoked vehicles, because one HashedId8 identifying an Authorization Ticket measures 8 bytes and Authorization Tickets are appended to the list as they are revoked. For this simulation we used an adaptive technique to the Bloom Filter CRL approach, since the allocated space for a BF is fixed, we increase it step by step as the filter fills up. With this approach, the space of the BF is allocated when needed, minimizing the waste of bytes. When a new BF is created, the content of the previous one is transferred to the new one, in this way the new BF represents the whole CRL.

Once again in this graph we can observe the relationship between the FPP and the compression ratio (R). Conservative values of compression ratio (R less than 6.5) would guarantee a low FPP and consequently fewer backup certificates needed.

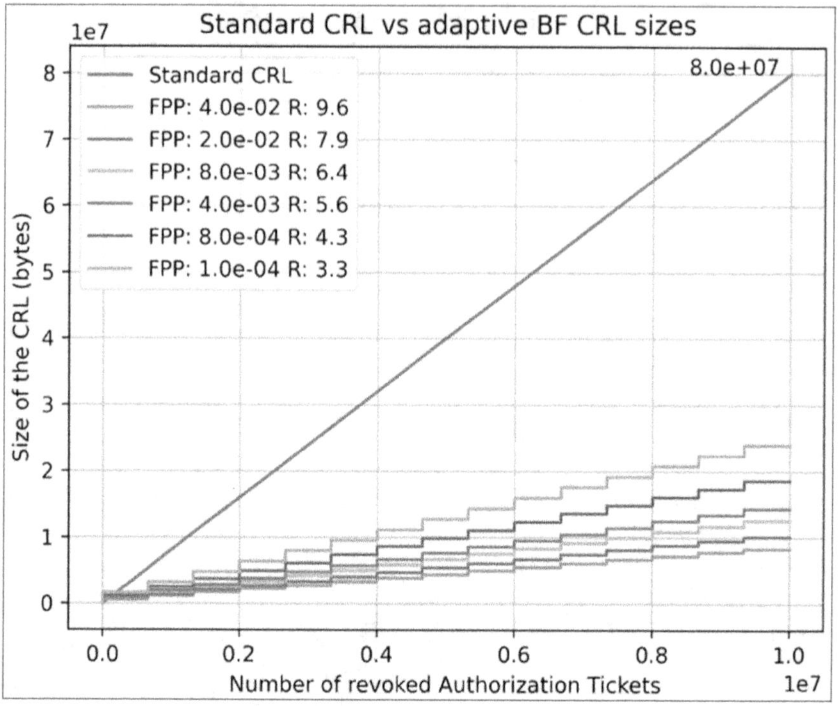

Figure 16. Standard CRL vs adaptive Bloom Filter sizes comparison for different input parameters of the BF.

Attacks on Authorization Tickets Tracking

Each V2X message sent by a connected vehicle must be signed with an Authorization Ticket (AT). These ATs are anonymous and do not reveal any information about the vehicle or the driver. The value that ATs brings is that a V2X spoofer could not recognize the driving (and living) patterns of a targeted person.

An advanced spoofer could try to track a vehicle that sends the V2X messages signed with the same AT. This spoofer would obtain the trajectory of the car and this information could be cross-correlated with additional data obtained by other means that would reveal the identity of the driver. For example, the starting point of the trajectory would be the driver's home address and the end point would be the work place. The Authorization Authority solves this problem by assigning periodically a set of ATs to each connected vehicle. In this way, the connected vehicle can change the AT at any time and break the tracking line that the spoofer has been following.

The spoofer could go further and try to reconstruct the trajectory of a vehicle that has signed the V2X messages with different ATs. This is not an impossible task, since the connected vehicle sends the V2X messages in periods ranging from 100 to 1000 milliseconds. That means that a vehicle running at 39 km/h will only advance a few meters during the time in between two messages sent (1 meter if the time is

100 milliseconds and 10 meter if the time is 1000 milliseconds). This proximity makes it easy to relate two V2X messages with different ATs to the same vehicle, especially when there are not more vehicles around them.

A most extreme case would be when the spoofer is a trained ML algorithm that considers all the information available in the V2X messages to decide that two V2X messages with different AT belong to the same vehicle. This is the problem that is tackled in the present section and that we will try to mitigate.

Attack Mitigation with the Authorization Ticket Scheduler

This section proposes an effective AT scheduler architecture that decides the best moment to change the AT of a target vehicle to avoid being tracked by an attacker. The decision is made by evaluating how easily it is to track the car using the information contained in its V2X messages. The scheduler is designed to be placed in the vehicle's anti-hacking device, and it takes into account a buffer of old V2X messages from both the owner's car and the surrounding vehicles at a given time and place.

More precisely, a tracker tries to associate each V2X message sent by the target car with a subgroup of old messages that were sent by the same vehicle. Although the length of this subgroup can be variable, in practice, the tracker only needs to link the new message with the latest message sent by the target car in order to track it. The system stores a buffer of old V2X messages coming from the surrounding vehicles and the target car itself. The system is trained to discern which of those old messages correspond to the same target vehicle. This way, it is possible to evaluate periodically how well the car can be tracked by an external spoofer. If the target messages are associated with the correct subgroup with a high score, it means that the car can be easily tracked.

The AT scheduler is composed of three main modules (Figure 17):

- A **V2X message candidates selector**, that selects a group of N old messages that are most likely to belong to the target vehicle.
- A **V2X message tracking scorer**, that associates the incoming message with a subgroup of the N candidates with a certain score.
- An **AT change decision engine**, that decides the best moment to change the AT of the V2X message according to a buffer of tracker scores, the number of remaining ATs and the time to get a new pack of ATs.

Each of the modules are further explained in the following subsections.

The AT scheduler architecture has been designed following a modular approach: a pipeline of two independent blocks that work simultaneously. The first block receives the tracking messages of the surrounding vehicles and from the car itself and uses a tracking algorithm to produce a score that asses the tracking difficulty,

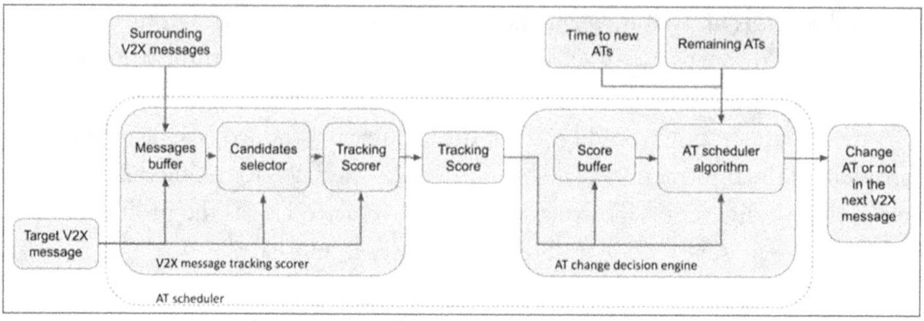

Figure 17. AT Scheduler Architecture.

simulating a spoofer. These scores are used in the AT change block. This block decides when is the best time to change the AT given the scores in time, the remaining valid ATs in the car and the time to receive new ATs.

AT Scheduler Modules

This section provides an in-depth description of the three main modules that form the AT scheduler. All of the modules are used once for each incoming V2X message from the target car.

V2X Message Candidate Selector

The candidate selector module is responsible for providing a batch of passed messages (candidates) that have chances to belong to the same car as the incoming message (target), as well as some features that can be used by the tracker to link them. To select the candidates, the system stores a list of old V2X messages that were sent by the surrounding cars along with those that were sent by the target car. This list is periodically updated to keep only the last M messages, so older messages are discarded to maintain a manageable amount of data.

The candidate selector takes as input all the messages in the buffer. As all ATs should be unique, if the incoming AT matches one of the old messages it can be directly linked without further analysis. If not, the module computes some features of the candidates that can be useful to link them to the target. All the features are assumed to be computed using information that is available in the V2X messages. These features may include, but are not limited to:

- The time difference between the candidate and the target.
- The variation between the target's position coordinates and the expected position that the candidate should have at the time of the target.
- The variation between the target's velocity and acceleration components and the candidate's expected values of those components at the time of the target.

Finally, the candidates are sorted by the values of their features and the selector returns the N candidates with the smallest variations so that they can be analysed by the tracker.

V2X Message Tracking Scorer

Given a set of candidates and their features, the tracking scorer works as a regression algorithm that chooses to which candidate a target message is linked, meaning that the two messages were sent by the same vehicle.

The proposed method to implement the tracking scorer is a Random Forest [4], which is an ensemble learning method consisting of a fixed number of decision trees. Each decision tree evaluates a random subset of the available features with specific thresholds to decide whether the candidate messages belong to the same vehicle as the target message. For each candidate, each tree outputs a binary value $y \in \{0, 1\}$ and the Random Forest outputs a single score $Y \in [0, 1]$ representing the average prediction of the decision trees.

One of the main advantages of Random Forests is that they are less prone to overfitting than a single deep decision tree, thanks to the creation of random subsets of features. Ensemble methods can produce more accurate results than any of their individual predictions while reducing the variance of the outcomes. Random Forests are also less computationally demanding than other common AI classification systems, such as Neural Networks.

The Random Forest is trained using the synthetic dataset of V2X messages created from an open data vehicular mobility dataset [5]. The produced dataset contains V2X messages of multiple cars with random periods of 100 ± 50 ms, each one containing the following variables:

- Message timestamp.
- Car id.
- x and y coordinates of vehicle position.
- x and y components of vehicle velocity and acceleration.
- Module of the vehicle velocity and acceleration.

AT Change Decision Engine

The AT change decision algorithm approach is modelled as a well-known mathematical problem known as optimal stopping, also called the marriage problem, secretary problem or best-choice problem. Selecting the best time to change the AT is crucial to minimize the spoofing tracking capabilities. In the optimal stopping scenario, a decision-maker observes inputs that evolve in time and involve some randomness and decides when is the best moment to perform an action given the known inputs. A time limit to make the decision also needs to be set, otherwise the decision-maker would be observing the inputs indefinitely looking for the best moment to perform the action. In this system, the decision-maker is the AT change decision algorithm, the inputs are the scores given by the V2X tracker scorer, the action is to change the AT and the time limit is a function defined below.

It happens that the number of actions (changing the AT) that can be performed depends on the number of non-used AT that the vehicle still has. Once all the AT of the vehicle have been used, not more AT changes will be possible until the vehicle receives new AT from the Authentication Authority. To distribute the AT equally on time, the time limit to change the AT has been defined as a function of the remaining non-used AT in the vehicle and the time left to get new AT from the Authentication Authority:

$$Time\ limit = \frac{time\ to\ new\ ATs}{Remaining\ ATs}$$

It is important to highlight that the time limit will be updated after each time that the AT is changed. It has been proven [6] that the best decision can be made with a probability greater than 36%. The strategy to be followed is to spend the 37% of the available (time limit) observing the tracking scores and get the maximum value of these scores. Then check the following scores until getting a value that is greater than the maximum value obtained during the observation time. If any greater value appears during the rest of the available time and the time limit is reached, the system forces an AT change.

OBU Hardware Securization

General Architecture

The On Board Unit is composed by a combination of three sets of hardware: the main board module, the V2X module, and the LTE module. The V2X module includes an independent processor, a Hardware Secure Module (HSM), a GNSS receiver and V2X transceiver. The connections between the different elements of the V2X module are secured using anti-tampering hardware and software techniques. Figure 18 shows the different blocks forming the OBU, the connection interfaces between the different elements and the elements that have been secured.

The OBU modular platform is designed to provide services related to communication, information and entertainment to the vehicle and at the same time be adaptable to the customer easily in order to perform demos and field tests. This modular platform is named CarCom.

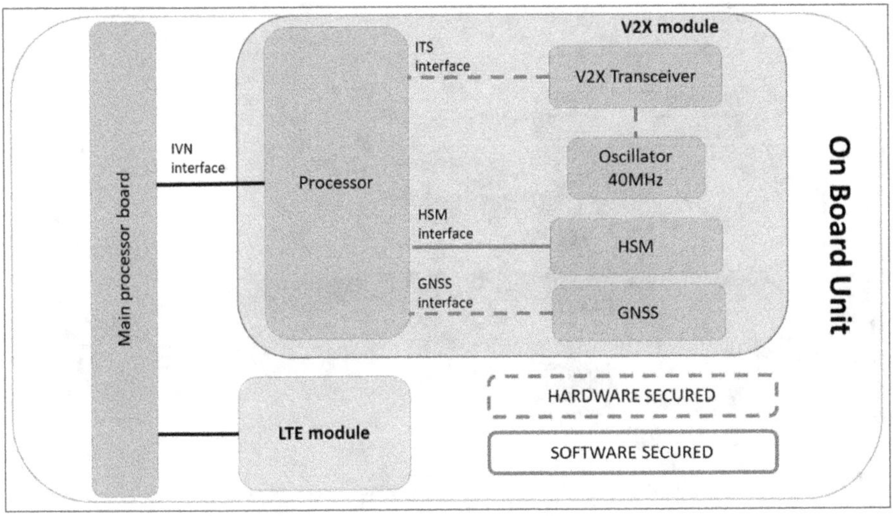

Figure 18. High level scheme of OBU system blocks.

The OBU has been designed and developed as a modular design and it is composed of several boards connected between them using a high density edge connector. The Main Board is responsible to allocate up to 4 daughter boards called STICKS or modules.

The following table shows the main technical features of the OBU modular platform:

Operational temperature	−40°C to +85°C
Voltage	9–18 Vdc
Power Consumption	<25W (peak) <12W (typ) <24mW (standby)
Core architecture	Dual ARM Cortex A7 up to 600 MHz, with MMU, FPU and NEON support + Cortex M3
RAM Memory	512MB DDR3L SDRAM
Flash Memory	1GB NAND Flash
WIFI	802.11a/b/g/n/ac MIMO RSDB (Real Simultaneous Dual Band)
Bluetooth	v4.2
Ethernet	1000Base-T1
Audio interface	A2B Transceiver AD2425W
Other interfaces	Slot SD card
Stick 1 option	DSRC V2X with iMX6UL + 2 SAF5400
Stick 2 option	Cellular 5G modem Sierra Wireless em7565

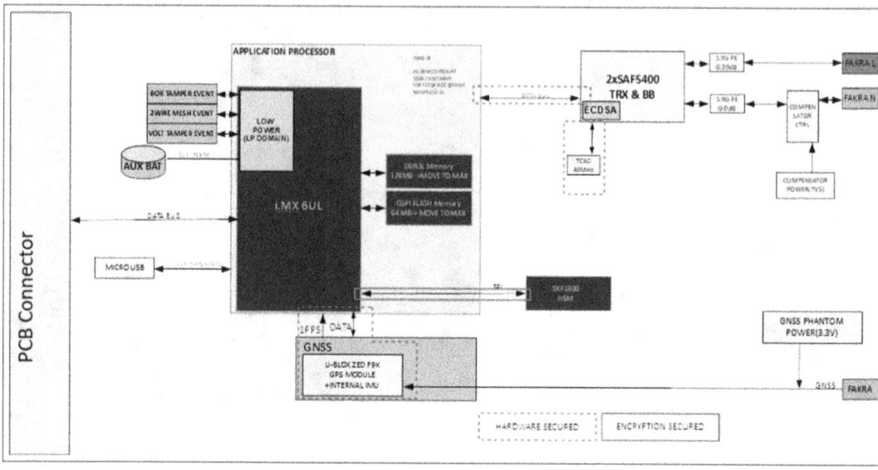

Figure 19. V2X module hardware blocks.

CarCom power supply adopts multiple DC/DC converters in the 20W~30W range and few low dropout voltage regulators (LDOs).

The V2X module (Figure 19) is a dual channel 802.11p secured system based in NXP's Chipset. It contains a single-core i.MX6 UltraLite processor, dual SAF5400 transceivers with an internal ECDSA, external secure element HSM SXF1800, an integrated GNSS module and tampered proof PCB, resulting in a secured platform designed to provide secure communication functionalities.

The i.MX6 UL has a 32 bit processor with 1300 DMIP/core featuring an advanced implementation of a single ARM® Cortex®-A7 core, which operates at speeds up to 696MHz. This processor has 256Mbytes of RAM and 256Mbytes of FLASH memory.

The i.MX6 UL host processor acts as the master of the system, booting the Operating System and controlling the SAF5400 chipset. Among others, the host processor is responsible for running the V2X communication stack. In terms of V2X communication stack, the imx6ul hosts the V2X-Com module and lower software layers like BSP or secure interface. With regards to radio frequency, the module provides 2 V2X antennas one of which is direct and the other uses a compensated antenna. It also includes a GPS active antenna.

Figure 20 shows a photo of the V2X module with indications of the different hardware elements mounted on the board.

Software NXP BSP

The software for the CARAMEL project builds on top of the BSP (Board Support Package) that NXP provides for NXP's V2X evaluation board (this board is Named Roadlink EVK 2.0). This BSP is based on the Yocto project.

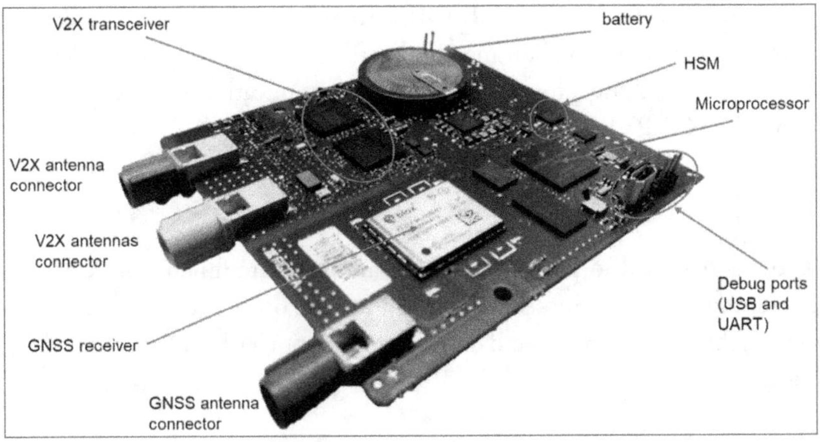

Figure 20. V2X module hardware.

The Yocto Project is an open-source collaboration project that helps developers to create custom Linux-based systems for embedded products. The project provides a flexible set of tools that can be used to create tailored Linux images for embedded devices by means of meta-data that describes how to construct the Linux distribution.

Yocto meta-data is defined in files called recipes that contain a list of settings and tasks (instructions) for building packages. In addition, recipes describe dependencies for libraries or other recipes, as well as configuration and compilation options. Layers are collections of related recipes, bundling related metadata to customize your build.

Layers and the recipes in them contained are parsed by the Yocto build tool named bitbake. Bitbake is very flexible and enables to build individual recipes or even specific tasks within a given recipe.

CARAMEL adds its own layer to the Yocto build system provided by NXP in order to add support for the project.

In terms of V2X communication stack, the BSP from NXP only contains the lower V2X software layers required to access and evaluate the hardware. The V2XCom module, explained earlier, provides the ETSI G5 protocol stack (GeoNetworking, BTP and Facilities).

The reference V2X BSP is provided as three sets of Yocto layers:

- A first set is formed by NXP's generic upstream Yocto releases for the i.MX6 platform based on the 4.1.15 Linux kernel and for the i.MX8 platform based on the 5.4 Linux kernel.
- A second set is formed by the V2X specific layers, that customize and alter the upstream Yocto release with V2X specific applications, forming a V2X reference image where the user can base its own designs.

- A third set is formed by CARAMEL specific layer. This layer adds the low level information for booting the board and supporting all the contained peripherals. It also provides the V2X communication stack layers not provided by NXP's BSP.

Technical Safety Specifications

The OBU is designed to provide hardware and software functionalities to ensure secure communication. The security design focuses on the communication interfaces inside the OBU, to avoid and detect manipulation either by physical or logical methods.

STRIDE Model

STRIDE is a threat modelling technique where potential threats, vulnerabilities, or the absence of safeguards of a system can be identified and enumerated. Once threats are identified we will decide what defences or countermeasures need to be included.

STRIDE is an acronym of Spoofing, Tampering, Repudiation, Information Disclosure, Denial of Service and Elevation of Privilege. These represents the six major categories of threat, which are:

1. **Spoofing** involves gaining access to a system violating authentication.
2. **Tampering** is the unauthorized modification of the system, which violates integrity.
3. **Repudiation** is the ability of users to claim they didn't do something or making it impossible to link an action back to you, which violates non-repudiation.
4. **Information disclosure** is the unwanted exposure of confidential and sensitive information.
5. **Denial of service** involves exhausting resources required to offer services, and makes a system or application unavailable.
6. **Elevation of privilege** occurs when a user assumes the identity of a privileged user to do what they are unauthorized to do.

Potential Threads

We use STRIDE modelling tools to identify potential threats, keeping in mind the OBU architecture and interfaces, then we walk through each STRIDE threat to identify potential OBU attacks. Following STRIDE, the potential threats for tamper attack of vehicle's OBU that have been considered in this project are identified in Table 1.

Table 1. Potential Threats for attack of vehicle's OBU considered in CARAMEL.

Title	Description	Method	STRIDE
Enclosure manipulation	The attacker wants to get access inside the OBU to make modifications	Open metal lid with a screwdriver	Tampering
Hardware attack on the ITS interface	The attacker uses tampered V2X messages to cause safety hazardous situations	Physically connecting to ITS signals or even replace V2X transceiver.	Spoofing Tampering
Software tamper attack on the ITS interface	The attacker uses malicious software on the V2X front end to track ITS stations or to send rogue messages on the ITS network	Physically connect to the OBU and attempt to reflash the chip to inject malicious software	Tampering & Elevation of privilege
Clock fault injection attack on the ITS interface	The attacker manipulates front end's clock to generate malfunctions or break security in the ITS interface	Physically connect to the V2X transceiver and manipulate system clock in order to induce a failure to gain access to the restricted area	Tampering, Denial of Service & Elevation of Privilege
Software tamper attack on the main processor	The attacker uses malicious software on the main processor to cause safety hazardous situations.	Physically connect to the OBU and attempt to reflash the CPU to inject malicious software	Tampering & Elevation of privilege
Clock fault injection attack on the main processor	The attacker manipulates main processor's clock to generate malfunctions or break security	Physically connect to the CPU and manipulate system clock in order to induce a failure to gain access to the restricted area	Tampering, Denial of Service & Elevation of Privilege
Voltage fault injection	The attacker manipulates power supply to generate malfunctions or break security	Physically connect to the main voltage and manipulate in order to induce a failure to gain access to the restricted area	Tampering, Denial of Service & Elevation of Privilege

(Continued)

Table 1. Continued

Title	Description	Method	STRIDE
Temperature fault injection	The attacker manipulates environmental temperature to generate malfunctions or break security.	Manipulate temperature in order to induce a failure to gain access to the restricted area	Tampering, Denial of Service & Elevation of Privilege
Eavesdropping main processor data signals	The attacker eavesdrop communication of the main processor memory to get secrets	Physically connect to CPU signals	Information disclosure
Hardware attack on the HSM interface	The attacker uses tampered HSM messages to cause safety hazardous situations and to get privileges.	Physically connecting to HSM signals or even replace HSM.	Spoofing Tampering
Software tamper attack on the HSM interface	The attacker uses malicious software on the HSM to cause safety hazardous situations.	Physically connect to the OBU and attempt to reflash the HSM to inject malicious software	Tampering, Denial of Service & Elevation of Privilege
Hardware attack on the IVN interface	The attacker tamper IVN data to cause safety hazardous situations.	Physically connecting to IVN signals.	Spoofing & Tampering
Hardware attack on the GNSS interface	The attacker uses malicious geolocation data to cause safety hazardous situations.	Physically connecting to GNSS signals and attempting to inject malicious messages.	Spoofing & Tampering

Countermeasures and Anti-tamper Techniques

In hardware tampering attacks, the adversary actively interacts with the device and/or its components by, for instance, inducing deliberate faults into the computation and observing its result at the output. The severity of the tampering can range from just naive manipulation such as breaking a seal, to dangerous manipulation resulting in accessing privileged information. Therefore, tampering attacks are directed to a specific vehicle affecting privacy, and potentially, safety.

Since anti tamper techniques are not fool-proof, an "onion layered" security approach is necessary. Overlaid techniques provide more robust protection: the attacker must disable a protection layer before dealing with the next level of protection.

In order to comply with the security functional requirements of the CARAMEL project, several security layers have been applied, including hardware security and software security. The security techniques used are explained below:

Environmental sensors

Some kind of attacks can be done without physically opening the OBU, such as voltage fault injection attacks (on the main power supply) and temperature fault injection attacks (with climatic chamber).

The microprocessor used has temperature and voltage sensors embedded, as well as clock monitoring circuitry. With these sensors we are able to detect temperature, voltage and clock fault injection attacks.

External HSM also includes several sensors to prevent the device from being used outside the normal operation conditions. Operating outside these conditions can be considered as an attack. In this case, the device performs a security reset to prevent a potential attack.

Open box detection switch

In order to avoid the attacker getting access into the OBU, we have implemented a switch which is able to detect box opening and triggers an alarm in the microprocessor (Figure 21).

Figure 21. Open box detection switch.

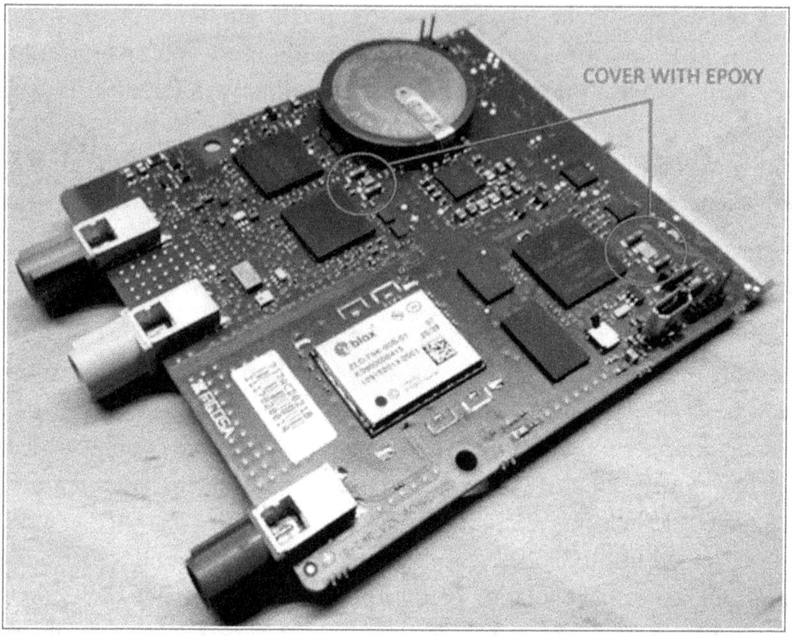

Figure 22. Clock oscillators covered with epoxy.

Coating covering sensible circuits

Clock oscillators are a common objective for fault injection attacks. In order to avoid that, clock oscillators will be covered with hard epoxy resin, together with two resistors (Figure 22). If epoxy is removed, resistors will be broken, and an alarm will be triggered.

Active wire-mesh protection

We design an active wire-mesh covering secure signals (Figure 23). Active wire-mesh consists of two differential signals: one attached to battery voltage and another one attached to GND. The two signals are traced in parallel covering the secure signals in order to avoid accessing these signals by drilling. If some of these wire-mesh differential signals are broken an alarm trigger is detected. Short circuit between the differential signals is also detected and triggers an alarm.

In the Figure 24 we show the secure signals between microprocessor and SAF5400, which are in green, and the wire-mesh used to cover these signals on the top layer (Figure 25).

Wire-mesh layout is made with an irregular pattern so that it will be difficult for the attacker to identify each of the differential lines. There is also another wire-mesh in an inner layer under the secure signals to protect them against access from the bottom layer. The same solution with wire-mesh is used to protect communication signals from the microprocessor to the GNSS engine. We use laser via and buried

Figure 23. Example of wire-mesh lines.

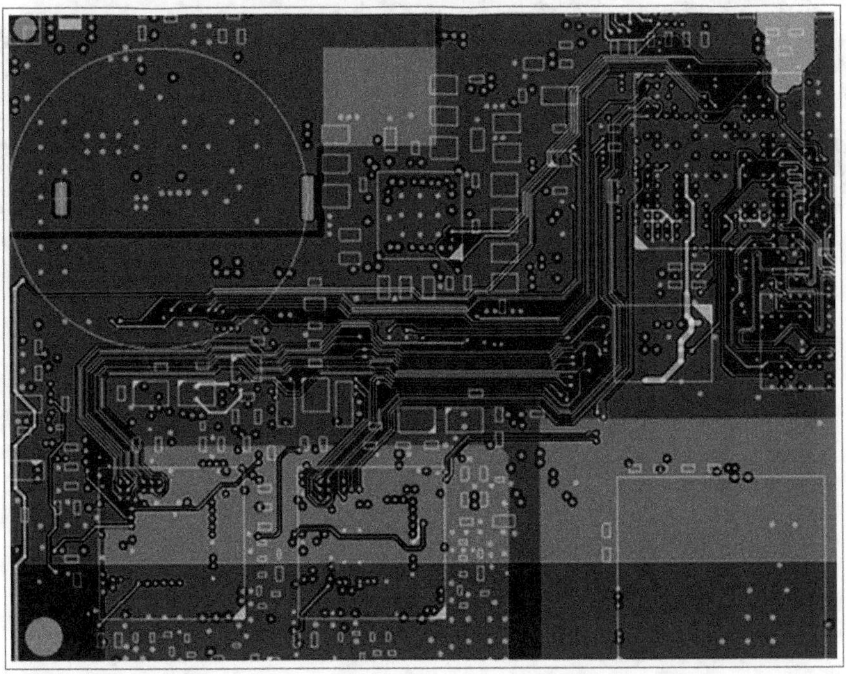

Figure 24. Secure signals between microprocessor and V2X transceiver.

via to connect wire-mesh to the microprocessor, which detects the attack to the wire-mesh (Figure 26). This is to restrict access to the wire-mesh from external layers.

Wire-mesh GNSS protection

An attacker could be able to open the GNSS metal lid to gain access to the interior of the GNSS module. In order to avoid this, we design a frame with wire-mesh made with vias, and then the frame wire-mesh is also connected to a cover which

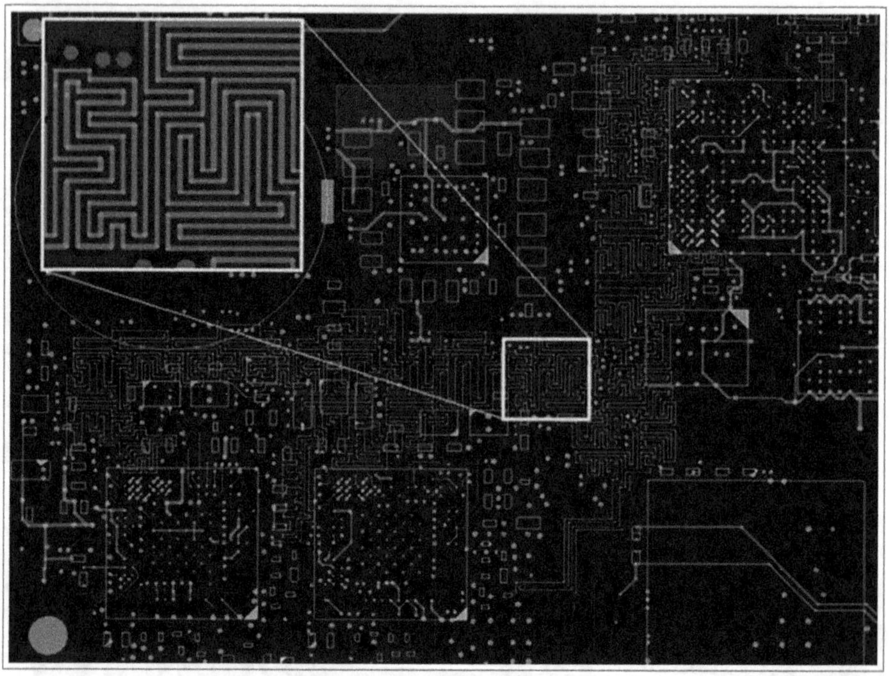

Figure 25. Wire-mesh used to cover secure signals from top layer.

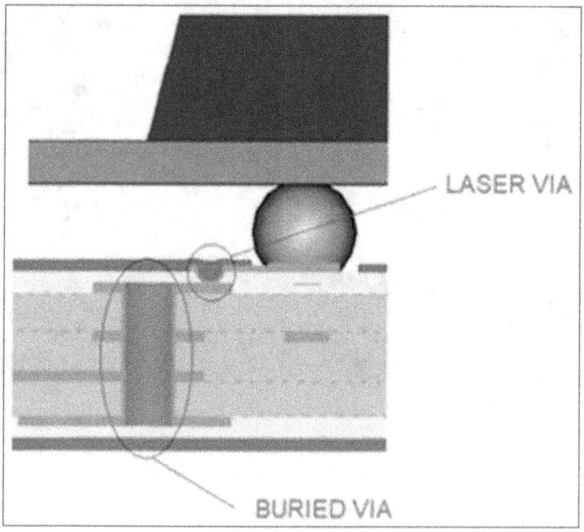

Figure 26. Connection of wire-mesh with microprocessor (transversal cut).

also has a wire-mesh (Figure 27). Then, if the cover or frame is detached or drilled an attack is detected.

Connection pads also have a guard ring of the opposite voltage level. This is because if the attacker wants to use conductive ink to get access to the pad, the

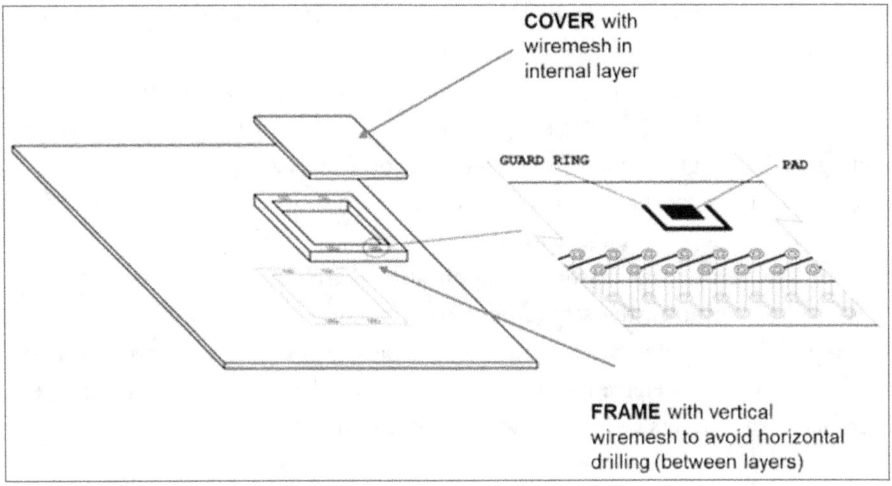

Figure 27. Illustration of GNSS metal lid protection.

ink will make a short circuit between the pad and the guard ring and the attack is detected.

Mutual authentication

The HSM is being protected against lifting and using in an unintended environment by requiring mutual authentication at each start-up. It requires possession of secret keys, which are re-programmable and initially provisioned by the manufacturer. The host processor should store the keys safe and secure. If the keys are lost, the device becomes unusable.

Data encryption

A secure channel is used between the host processor and external HSM to protect the authenticity, integrity, and confidentiality of the data transferred in both directions. The implementation is according to GlobalPlatform Secure Channel Protocol '03' (SCP03). The protection ensures integrity and confidentiality of messages exchanged between the HSM and the host processor.

Secure boot

The host processor includes the ability to perform a secure boot. This feature uses a combination of hardware and software together with a public key to protect the system from executing unauthorized programs.

Before secure boot allows a user's image to execute, the image must be signed. The signing process is done during the image build process by the private key holder and the signatures are then included as part of the final Program Image. Then, during the secure boot, the Read Only Memory (ROM) verifies the signatures using the public keys included in the Program Image.

In addition to supporting digital signature verification to authenticate Program Images, Encrypted boot is also supported. Encrypted boot can be used to prevent cloning of the Program Image directly off the boot device.

V2X transceiver and HSM will be booted also in secure mode. After reset, based on the boot mode, the image is loaded followed by an authentication and decryption stage. When it is successful, the application starts.

With these features, the secure boot component of the ROM protects against the potential threat of attackers modifying areas of code or data in programmable memory to make it behave in an incorrect manner. The secure boot also prevents attempts to gain access to features which should not be available.

Trusted execution environment

The secure microprocessor architecture provides a trusted execution environment for security-critical software. Software running in this environment is protected against attacks from potentially compromised platform software, including applications, services, drivers, and even the operating system itself. The trusted execution hardware protects the confidentiality and integrity of both security services and sensitive data. Furthermore, security services cannot be starved of access to processor resources or hijacked by uncontrolled interrupts. Trusted execution environment allows security-critical software to coexist with a rich platform software environment.

The trusted environment architecture is capable of distinguishing between four different code execution modes:

- Code executing in Normal World mode:
  - Code running in kernel mode (also called supervisor mode or privileged mode).
  - Code running in user mode.
- Code executing in Trusted Secure World mode:
  - Code running in Trusted kernel mode (also called supervisor mode or privileged mode).
  - Code running in Trusted user mode.

Countermeasures and potential threats mitigation

Table 2 shows the different countermeasures which have been implemented in the design, and the mitigation of the potential threats affected from each countermeasure.

Table 2. OBU protection levels implemented in CARAMEL and their benefits.

	Countermeasures \ Potential Threats	Enclosure manipulation	ITS interface HW attacks	ITS interface SW attacks	Clock fault injection	CPU HW attack	CPU SW attack	Voltage fault injection	Temperature fault injection	HSM interface HW attack	HSM interface SW attack	IVN interface HW attack	GNSS interface HW attack
HARDWARE	Environmental sensors				■			■	■				
HARDWARE	Opening enclosure detection	■	■	■	■	■	■	■	■	■	■	■	■
HARDWARE	Coating covering sensible circuits, with self-destructive components to avoid coating removal				■								
HARDWARE	Active wire-mesh protection for critical elements and signals		■		■	■							■
SOFTWARE	Mutual authentication		■			■				■			
SOFTWARE	Data encryption					■	■			■	■	■	
SOFTWARE	Secure boot			■			■				■		
SOFTWARE	Application processor trusted execution environment			■			■						

Final Testbed and Demostration

The final architecture of the demostration of this use case is shown in Figure 28 and consists of the following components:

- Devices deployed in vehicles: Anti-hacking devices and OBUs with only LTE-Uu radio interface or LTU-Uu and 802.11p radio interfaces.
- IEEE 802.11p Roadside Units (RSU) network deployed in the area of the demonstration.
- LTE network with small cells deployed in the area of the demonstration, the dRax to control the small cells, and the virtual Evolved Packet Core (vEPC) running in the Multi-access Edge Computing (MEC).
- Ethernet network to connect small cells and RSUs with the MEC.MEC.
- Public Key Infrastructure (PKI) servers.
- Backend.

With this use case we demonstrate two operations:

- The interoperability between different radio technologies.
- The attack on the V2X Message Transmission.

The interoperability between different radio technologies is necessary when vehicles in the same region are using On-Board Units (OBUs) equipped with

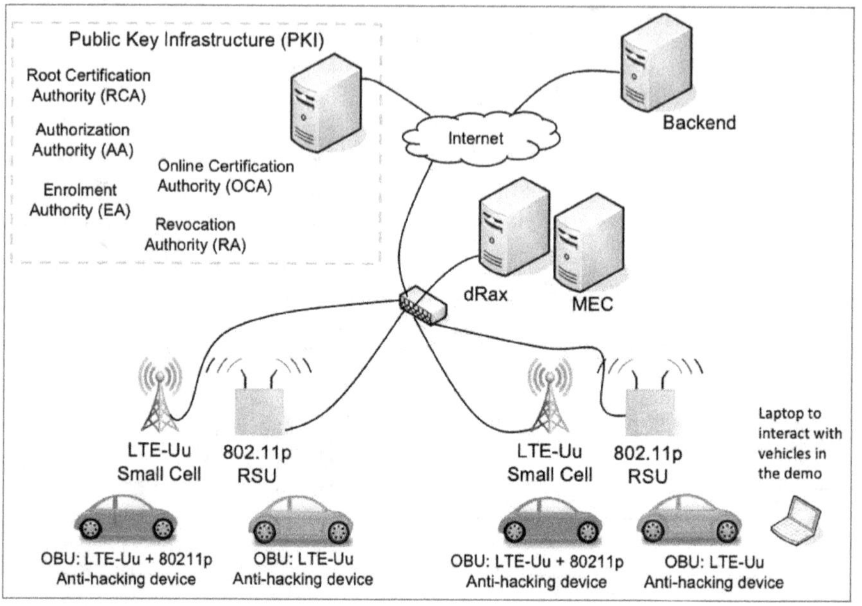

Figure 28. Global components of the testbed for attack on V2X message transmission.

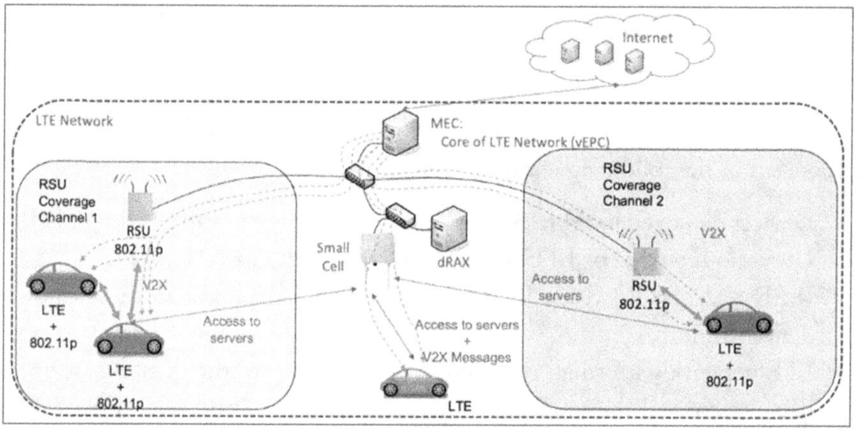

Figure 29. Forwarding between different radio technologies and/or regions of interest.

different standardized access technologies. Currently, it is possible to have OBUs with a single cellular connection (LTE-Uu), or others with this cellular connection plus a Vehicle-to-Vehicle (V2V) technology such as IEEE 802.11p, LTE-PC5, the new NR- PC5 or the not yet standardized IEEE 802.11bd.

As it has been described previously, CARAMEL's testbed consists of two types of vehicles, the "LTE-Uu only" and the "LTE-Uu + 802.11p". To perform the interoperability, we deploy a fixed infrastructure (Figure 29) consisting of an 802.11p RSUs network and a small private LTE network, both connected to a Multi-access

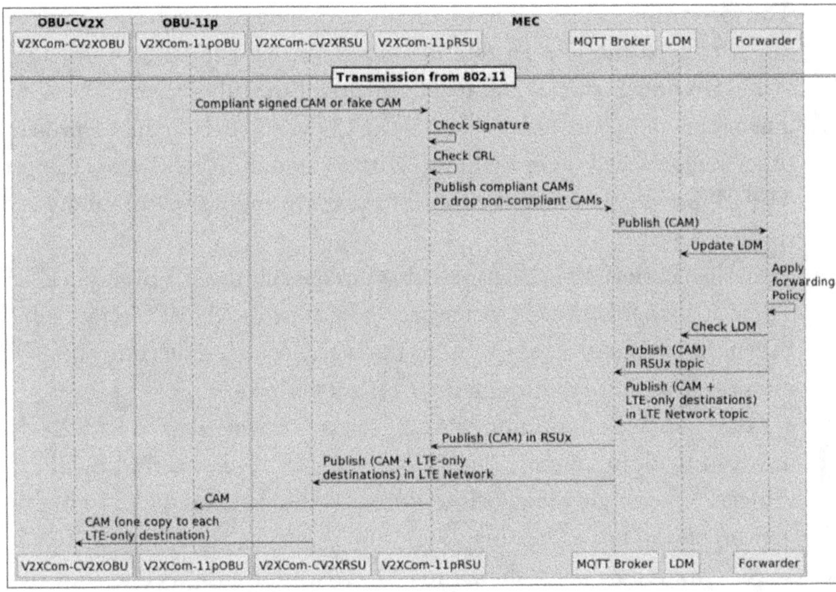

Figure 30. Workflow of use case "Interoperability between different radio technologies" when the transmission is initiated from a vehicle provided with 802.11p and LTE-Uu interfaces.

Edge Computing (MEC) which, using different kinds of policies, forwards messages from a radio technology to the other.

Figure 30 shows the workflow of a use case where CAM messages are transmitted through the 802.11p interface, a real V2V communication interface, which enables neighbouring vehicles under coverage to receive these messages. Nevertheless, vehicles "LTE-Uu only" or those that are far away do not receive them. The demonstration consists in doing that these vehicles also receive the messages using the forwarding mechanism through the MEC and the radio infrastructure.

The demonstration procedure workflow is:

1. One 802.11p vehicle transmits a compliant CAM which is received by the surrounding 802.11p vehicles and by the 802.11p RSU covering the area, for example RSU1 (note that in the testbed we have two RSUs: RSU1 and RSU2, covering two regions).
2. RSU1 forwards the message to its controller V2XCom module running in the MEC, which checks its validity.
3. If the message is non-compliant, it is dropped.
4. If the message is valid, it is sent to the "Forwarder" module, also in the MEC, using the MQTT broker.
5. The forwarder module, updates the LDM database of the MEC and checks the current forwarding policy.

6. The forwarding policy for our testbed is configured in such a way that messages received through an 802.11p interface, have to be forwarded to all "LTE-Uu only" vehicles and to all vehicles of the other region.
7. Therefore, the "Forwarder" checks the LDM to consult the identifiers of other regions' RSUs (in our testbed there is only RSU2) and, using the MQTT broker, forwards the CAM message to the V2XCom module controlling RSU2.
8. Also, the "Forwarder" checks the LDM to consult the "LTE-only" vehicles. Then, it forwards the CAM message and a list of all "LTE-only" vehicles that need to receive a copy of it, to the V2XCom module controlling V2X messages transmitted through the LTE network.
9. V2XCom module controlling RSU2 forwards the message to RSU2, which in turn broadcasts the message in Region 2 and it is received by 802.11p vehicles. One single message is sent from RSU2 because 802.11p is able to perform broadcast addressing.
10. V2XCom module controlling the LTE network transmits one copy of the message to each of the vehicles on the list. This transmission is done using the IP addresses of "LTE-Uu" only vehicles, which is stored in the LDM. We need to transmit one copy of the message per vehicle because the majority of LTE-Uu operators do not allow multicast transmissions.
11. The testbed user can check that all vehicles see the other ones using a computer connected to the backend LDM server.

The attack on the V2X message transmission consists of two different kinds of attacks:

- A malicious attacker transmits fake V2X messages. This is demonstrated in the testbed.
- A malicious attacker tries to track a specific vehicle. This is demonstrated by simulation.

CARAMEL addresses 5 types of fake V2X messages which are detected in the OBU and/or in the MEC:

- Non-Signed messages: When this event is detected the message is dropped.
- Messages signed with a non-valid certificate: This is the case where the certificate used to sign messages is not issued by a valid Certification Authority. When this event is detected, the message is dropped, and an alarm is triggered.
- Non-authorized messages: This is the case where an OBU of a "passenger car" transmits V2X messages specifying that its vehicle type is an "emergency vehicle". The Authorization Ticket of this car has been issued in such a way

that the receiver detects the anomaly. When this event is detected, the message is dropped, and an alarm is triggered.

- Replayed messages: This is the case where an OBU captures a message transmitted by another OBU and retransmits it. A third receiver may receive both copies. When this event is detected, the message is dropped. It is also possible that the receiver only receives the replayed message. This case only represents a security problem if the message has been modified, which will be detected by the digital signature. Additionally, if the replayed message is received later than a threshold delay time, the message is dropped. An alarm is not triggered because this replayed message does not present any security problem, and it could also be replayed by the fixed infrastructure, which is, in fact, a compliant action.
- Messages signed with a revoked certificate: When this event is detected the message is dropped and an alarm is triggered.

In all previous cases, whenever an alarm is triggered, the backend receives a notification. The backend monitors alarms and performs statistics for management purposes.

The case of a malicious attacker trying to track a specific vehicle by sniffing its transmitted messages represents a passive attack. CARAMEL has developed an algorithm that computes when the best time is to change the vehicle's AT, trying to mimetize itself among multiple neighbor vehicles. To show how this algorithm works requires a relatively high number of vehicles, and it cannot be demonstrated in the testbed, for this reason it was demonstrated by simulation. In any case, the software to implement this algorithm has been integrated inside the anti-hacking device, but it is not being executed during the demonstration.

As an example, Figure 31 shows the workflow when an attacker vehicle transmits non-signed messages or signed with a non-valid certificate:

1. The attacking vehicle transmits a message signed with a certificate not generated by the Authorization Authority of the Caramel PKI system.
2. The receiver device, either an OBU or the MEC, performs two basic anti-reply operations which are checking if the message has already been received or if it is too delayed. As in this case the message is not replayed, the procedure jumps to the next step.
3. The receiver device checks if the certificate is present in the Certificate Revocation List (CRL). As this certificate is false, we assume that it is not present in the list.
4. The receiver checks the authenticity of the digital signature and the validity of the attached certificate. In the OBU, the attached certificate validation requires the assistance of the HSM where the Root Authority certificate, with

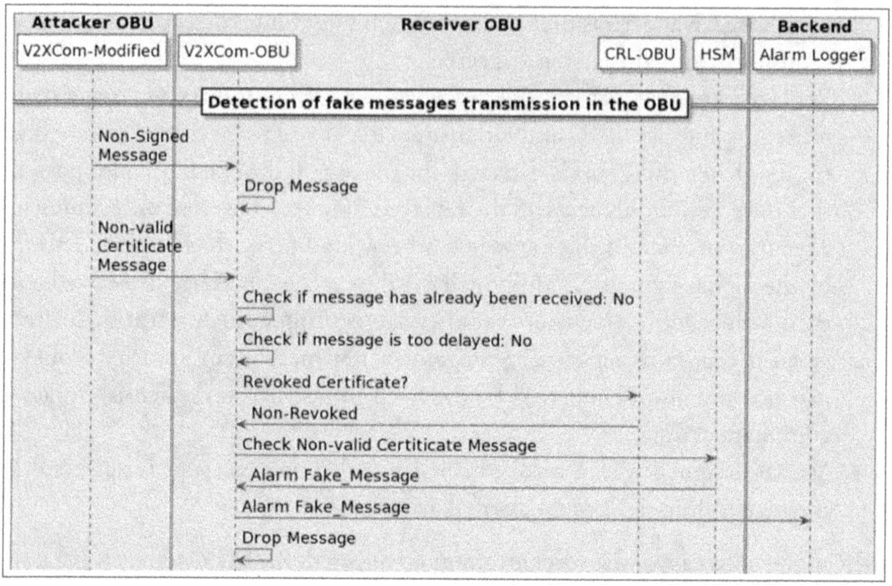

Figure 31. Workflow of use case "Interoperability between different radio technologies" when the transmission is initiated from a vehicle only provided with LTE-Uu interface.

which it is signed, is stored. In the MEC, as there is no HSM, this operation is uniquely done by the V2XCom-MEC module.

5. In this use case, the previous operation returns that the certificate is not valid and an alarm is triggered for statistical purposes to the backend.
6. The message is dropped.

Conclusion and Future Work

Pillar 2 presents solutions for three hot topics about the basis on which Intelligent Transportation Systems (ITS) relay. These are the implementation of ITS security models defined by the ETSI, the co- existence of different radio technologies to interchange V2X messages and the veracity of the positions that vehicles are announcing.

i2CAT, by developing the components of which it is responsible for, has set the roadmap of its participation in future V2X projects and services. The main component that i2CAT has developed is the ETSI ITS protocol stack containing different V2X messages, BTP and GeoNetworking protocols, its interface with a PKI architecture to manage ATs and the integration with different kinds of HSM. In the market, there are two main consumers of such a protocol stack, the car manufacturers that deploy OBUs containing the protocol stack in their vehicles, and road infrastructure operators that require the stack to be executed in servers to provide

ITS services to drivers and administrations. In CARAMEL we have used the same version of the stack for both segments, in the NEXTIUM by Idneo's OBU, and in the MEC. As the deployment of the whole stack departing from zero is a huge task, we have relayed in the open-source Vanetza framework, modified and upgraded conveniently to be adapted to the specificities of CARAMEL's use cases. We have seen that commercial versions of this stack provided by Commsignia, Cohda Wireless or Lacroix are much more efficient on the OBU side since their stack has been programmed very specifically for automotive computers, which have small memory and small computation capabilities. Therefore, i2CAT's stack cannot compete in this segment. Nevertheless, these commercial stacks are closed and focused to high level functionalities, the programmer can not access to specific functions or change the parameters of the protocols, they are not flexible, and they are hard to be adapted to different scenarios. On the other hand, these are, precisely, the characteristics of the ITS protocol stack developed in CARAMEL which adapts very well when used in a MEC or other servers in the infrastructure. For instance, it can be used to build digital twins, traffic management centers, to develop any kind of ITS application. Moreover, it can be used in other types of environments, for instance to build small size OBUs, using Raspberry Pi platform, for bicycles or electric scooters.

The second module that i2CAT has developed is the system architecture that contains a MEC, connected to fixed infrastructure of different types of radio technologies, with the necessary software components to allow vehicles, using these different types of radio technologies, to communicate among themselves. In CARAMEL we have demonstrated the case of interoperability between the V2X native standard IEEE 802.11p and LTE that provides non-native V2X communication, having to use V2X messages over IP. These components are valuable for i2CAT's upcoming projects because radio technologies will continue evolving, and we need to adapt our systems to this evolution. i2CAT's immediate roadmap contains the plan to implement interoperability with a new radio infrastructure based in LTE-PC5 and, when available, also based in NR-V2X and 802.11bd. Additionally, the fact of having developed the necessary components to transmit messages over IP over LTE, allows it to switch from LTE to 5G with very little effort.

Another challenge of cybersecurity in V2V connected mobility is location spoofing. The latter attack aims to compromise the self-positioning ability of vehicles. A location spoofing attack attempts to fool a GNSS receiver by broadcasting false satellite signals, focused on resembling a set of normal satellite signals. These spoofed signals may be modified in such a way to cause the receiver to estimate its location even kms away from its actual position. The impact of this attack is more devastating if we consider a group of connected vehicles, which exchange their location measurements in order to coordinate their actions. Broadcasting falsified

GNSS positions, then severe traffic accidents are more likely to take place, injuring drivers, pedestrians, cars, etc. UPAT has developed the associated algorithmic mitigation and detection solution. The main task of this collaborative defense mechanism is to run dedicated cooperative AI solutions that work on the data transmitted to the leader ego vehicle. It is responsible to receive the measurements of the cluster's vehicles, identify and mitigate the possible attacks on the GNSS receivers and feed the re-estimated positions back to the involved vehicles. UPAT's module has been deployed within the CARLA-ROS simulated framework, using CARLA simulator to generate data from cars' sensors and ROS nodes as the data consumers. Within this approach, it is of future plan to evolve the interoperability of the defense mechanism, by integrating a realistic V2V communication simulator which models network delay. In addition, it will be explored how the current centralized implementation will be transformed to a distributed based solution, without the need of a central node, which enables scalability, lower computational and deployment cost, as well as mitigation and detection ability.

UCY developed the in-vehicle location spoofing attack detection solution. It uses a threshold-based approach (i.e., static threshold value that is selected during the training phase) for detecting suspicious deviations between the current GPS location and the GPS-free vehicle location estimate that relies on vehicle measurements and absolute vehicle location (e.g., through cellular network localization solutions). In the immediate future, the UCY team will explore Machine Learning techniques to obtain a dynamic threshold that is adapted to changing conditions. For instance, as the vehicle moves from a rural to a suburban to an urban environment the GPS location accuracy and precision degrades (because of increasingly obstructed satellites), while the accuracy of cellular-based methods that are used to estimate the GPS-free vehicle location is better due to increasing cell tower density. This dynamic threshold selection approach is expected to robustify the UCY attack detection solution, i.e., reduce false positives due to variable environmental conditions. Another research direction will be towards enhancing the solution not only to detect the location spoofing attack, but also to mitigate the attack by reconstructing the attacked GPS locations. To this end, denoising autoencoder techniques will be investigated to remove any bias introduced in the GPS locations as the result of an attack.

From NEXTIUM by Idneo side, as a solutions partner who developed software and hardware securitization methods to protect the OBU's hardware against tampering and attacks, an evolution for the hardware protection could be the usage of Physical Unclonable Functions (PUF) for the securisation. A PUF is a physical object that for a given input and conditions provides a physically defined "digital fingerprint" output. In semiconductors, due to the submicron manufacturing process variations, every transistor from the integrated circuit has slightly different physical properties that can be measured (transistor threshold voltages, gain factor, parasitic capacitances…). These variations are not controllable in the manufacturing process and using these inherent variations as inputs for certain algorithms, the silicon fingerprint is turned into a cryptographic key that is unique for that individual chip and is used as its root key.

The same principle can be used for protection against tamper attacks for the electronics of the OBU without using a battery. For example, if we were able to measure the capacitance between traces in the wire-mesh used for protection, these capacitances will be unique in each device due to the manufacturing process and could be used to extract a cryptographic key. If the device is modified in any way, even if the wire-mesh is bypassed, the inherent capacitances will be different, obtaining a different cryptographic key and then detecting a tamper attack.

References

[1] V. Kumar, J. Petit, and W. Whyte, "Binary Hash Tree based Certificate Access Management for Connected Vehicles," in Proceedings of the 10th ACM Conference on Security & Privacy in Wireless and Mobile Networks (ACM WiSec), Boston, USA, July 2017.

[2] ETSI TS 102 940 – V1.3.1 – Intelligent Transport Systems (ITS); Security; ITS communications security architecture and security management. ETSI, 2018.

[3] M. A. Simplicio Jr, E. L. Cominetti, H. K. Patil, J. E. Ricardini, and M. V. M. Silva, "ACPC: Efficient Revocation of Pseudonym Certificates using Activation Codes," Elsevier Ad Hoc Networks, July 2018.

[4] Breiman, L. Random Forests. Machine Learning 45, 5–32 (2001). https://doi.org/10.1023/A:1010933404324

[5] S. Uppoor, O. Trullols-Cruces, M. Fiore, J.M. Barcelo-Ordinas, "Generation and Analysis of a Large-scale Urban Vehicular Mobility Dataset", IEEE Transactions on Mobile Computing, Vol. 13, No. 5, May 2014.

[6] Theodore P. Hill. "Knowing When to Stop". American Scientist, vol. 97, no. 2, March-April 2009, p. 126. DOI: 10.1511/2009.77.126

DOI: 10.1561/9781638280613.ch3

Section 3

Electromobility

Unauthorized access and control of EVSE stations and firmware modifications should be prevented.

Introduction

In recent years, there has been an increase in demand for electric vehicles. Plug-in Electrical Vehicles (PEVs) are starting to appear on European roads, where the smart-grid capable Electric Vehicle Supply Equipment (EVSE) stations are used for charging. Cyberattacks at EVSEs can affect energy metering data and leak private information, or they can cause more dangerous situations like overcharging the large lithium batteries found in PEVs (since the Plug-in Electric Vehicle is negotiating with the charger) or disturb the operational flow of the electrical grid. It is abundantly clear that for the sector to experience sustained growth, the creation of a dependable and secure EV charging ecosystem (EVCS) is crucial. The usage of EVCS management systems (EVCSMS) can allow enhanced capabilities that could result in improvements to the EVCS. Although the implementation of EVCSMS can be advantageous for the adoption of EVs, security and management issues have

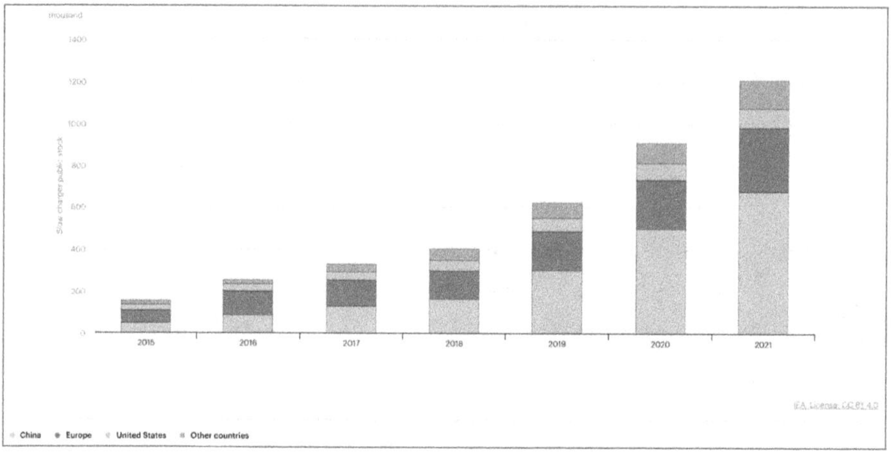

Figure 1. The increasing trend of publicly available chargers in different countries [2].

become a top worry for users and network operators due to the system's vulnerability and the vast variety of associated risks. Even though EVCSMS are created and offered by internationally renowned suppliers, it is unclear how a zero-day vulnerability could compromise a station's overall security.

Plug-in Electric Vehicles have shown significant promise in recent years, in part because of advances in electrical engineering technology, but also because of their potential to cut greenhouse gas emissions and lessen reliance on oil, which is consistent with society's growing environmental consciousness. The majority of the charging demand will likely be met by charging at home and at work, but in order to give consumers appropriate and practical coverage, the number of public chargers must increase nine-fold to over 15 million units in 2030 [1]. Figure 1 shows the growth of slow publicly available chargers, 2015–2021 in a few nations, with a rising trend over the past several years [2].

When it comes to lowering CO2 emissions in the transportation industry, plug-in EVs are seen as a viable solution. However, they come with a number of technical difficulties, such as the need for additional charging infrastructure. As the current power infrastructure cannot support the simultaneous charging of numerous electric vehicles, the installation of new charging stations could pose a serious difficulty to the Distribution System Operator (DSO). Plug-in EVs are considered a promising solution when it comes to reducing CO2 emissions in the transportation sector, but they are also accompanied by a series of technical challenges such as the additional charging infrastructure that must be installed in order to accommodate them. From the DSO perspective, the installation of a new charging station can become a severe problem as the current power infrastructure does not support simultaneous charging of large numbers of electric vehicles.

When building an EV charging infrastructure, the key objective is to make sure that all of the charge points are installed effectively and used in a secure manner. To avoid overwhelming the electrical network, charging for vehicles must be done simultaneously, safely, and affordably. By 2025, it is anticipated that there will be 190 million electric vehicles on the road, which translates to roughly 1,330 GW of power. This would be equivalent to the output of around 2,200 major power plants. Even while the popularity of EVs is growing, the exploitation of the infrastructure could have negative effects on the energy sustainability of the country and of Europe. In fact, the electric network would crash if too many EVs asked for electricity at once without any a priori dimensioning. Few research specifically address PEVs, despite the large number of studies that have been published about the general topic of smart grids. PEVs are regarded as distributed energy resources as well as being a component of the smart grid that contributes to the loads.

However, because neither the charging stations nor the DSO have implemented security mechanisms to identify and prevent security threats and attacks, or to mitigate the potential network disruptions caused by a breakdown (or hack) of the smart charging stations, this capability indirectly jeopardizes the dependability and security of the power network. Energy metering and payment, communication between the EV battery management system and the charge point, a communication mechanism between the Charging Point (CP) and a central management system, and finally the establishment of a communication channel between the CP and energy suppliers are all necessary components of the complex system that is smart charging (DSO, Transmission System Operators (TSO), smart girds, etc.). Due to the fact that the services are provided by several entities, the environment is made vulnerable to a variety of security threats on various levels by these complicated communication methods. Thus, this heterogeneity of the involved cyber-physical systems, makes the electrical system that is monitored and controlled from an ICT infrastructure difficult to coexist with the whole ecosystem. This necessitates the standardization of protocols and the implementation of two main interfaces, one for managing the system and the other for electricity. The ICT system is involved in the status, authorization, metering, and billing of the EV that interacts with the system in the event of smart charging.

State of Art/Innovation

Although several studies have been published on the general subject of smart grids, very few of them specifically focus on PEVs. PEVs are thought of as a part of smart grids, which distribute energy sources and sustain loads. The PEV distributed nodes are examined by [3], who highlights some of its potential flaws. The CAN bus used by PEVs is the subject of sniffing/replay attacks in [4], while [5] focuses on security features and IDS for usage inside the vehicle. For PEVs, typical cybersecurity attack scenarios that place the stability and integrity of the electrical distribution network at peril include False Data Injection Attack (FDIA), Advanced Metering Infrastructure (AMI) tampering, and data manipulation. These attack scenarios pose the risk of exposing user privacy information, resulting in erroneous payment transactions, altered smart meter measurements, serious damage to charging stations and autos.

In this regard, should there be a targeted cyber-attack on a sizable portion of the charging stations/EVs, with hackers, say, switching on and off charging every minute, the energy supply would not be able to keep up with this, and within a few minutes, a pan-European blackout would result. Modern smart charging solutions typically use platforms for cloud-based service operations and controllers that can fit many charging stations. Additionally, the algorithms at the heart of every smart charging system choose the best network capacity and charging needs while also taking current energy prices into account. All of these parts are enough to ensure the system runs without a hitch under ideal circumstances, but they must be supplemented with cutting-edge ICT modules that secure the smart charging PEV network and offer situational awareness through monitoring, sophisticated visualization, and intrusion detection methods.

By creating model-based and statistical intrusion detection approaches, CARAMEL's innovation has the objective to counter both known and zero-day threats. The former, are focused on detection of activities that significantly differ from system normal behavior, while the latter are based on modeling the characteristics of known assaults (using rule-based languages and state transition analysis toolkits). The fundamental frameworks for anomaly-based IDS will be machine learning, data mining, and statistical models. CARAMEL supports cyber-defense tools that can be configured to guard against intrusions using prevention, detection, and appropriate response techniques. The IDS consists of a number of strategically positioned probes on the smart chargers' data network. A centralized console assists in the proper management of the probes. All packets are analyzed by the software once data flow has reached the IDS.

The detection engine of CARAMEL can include conditional rules depending on the needs of the organization. The detection engine creates an event and logs it when a rule's requirements are satisfied. The smart grid and smart chargers cannot be protected just by an IDS deployment. The security analyst is primarily warned by the detection system when any malicious behavior is occurred at a specific moment. Thus, firewall interaction capabilities are added to the IDS probes. The smart controller network link includes the CARAMEL's virtual firewall applied in-line, allowing it to manage which packets travel via its two network interfaces. It can be launched with particular rules that only permit the propagation of the desired traffic. A real-time dashboard and database-driven analytics is offered by CARAMEL in addition to intrusion detection and prevention to visualize the current state, communication flows, network topology, and health status of the smart charging system. The raw data obtained from the IDS, system logs, or other network logs must be processed through a number of processes such data cleaning, data aggregation, and data transformation because they may contain errors or erroneous data. Through this process, redundant information is eliminated, and data is also properly formatted. Visual mapping is utilized to transform data objects into visual symbols once the network datasets are prepared. These specialized visual symbols will adjust to the requirement of presenting various data aspects for network alert, traffic, and attack data. Effective visual techniques are applied to ensure that human perception is included and in order to improve the ability to gather insights.

For the intrusion analysis requirements to be met, the visual symbols utilized for network representation are also made available in single or multiple perspectives. In order to assist the analyst in having an integrated and personalized picture of the events, some of the activities previously anticipated for the purposes of CARAMEL include the abilities of choosing, dragging, zooming, modifying color mapping and data mapping, as well as adjustment of the level of detail. By modifying the parameters of data processing procedures, choosing different visual mapping techniques, or tweaking visual representatives, intelligent and intuitive interactions are utilized to materialize the complete visualization pipeline. The dashboards that will be created will have multiple levels so that they may be used for system-wide, service-area, and per-device monitoring.

Threats/Problems Considered/Detected

It is possible for malicious actors to physically hack a charging point, for example with the intention of stealing energy. Another option is to attack and seize possession of a single charging point. Finally, a back office may possibly be attacked and partially taken over. Attacks like these cause communication breakdowns between a charger and the back office. For instance, the charging point may get inaccurate measurements or unexpected instructions. These counterfeit messages can be identified as anomalies since communication always follows a predetermined structure whose succession is predictable. There are numerous out-of-date chargers in use right now that are unable to handle the new communication protocols, even if they enable more secure data transmission between charging infrastructure and their back office. This security gap led to the need for remote detection methods in order to ensure the cyber security of these charge stations as well.

Attacks against Smart Charging have an effect on each element of the energy ecosystem, either directly or indirectly (charging infrastructure, TSO, DSO, etc.). The absence of security features in the charging stations for identifying and preventing potential assaults and threats can undermine the dependability and security of the entire energy supply network. A partial energy blackout might occur, for instance, if a potential attack vector in the smart charging infrastructure affects the DSO. Due to its complexity and the interactions between the parties involved, the smart charging infrastructure cybersecurity analysis identifies a number of potential attack routes. This communication architecture is vulnerable to a number of security risks on many levels. The metering and payment for energy, communication breaks between the EV battery management system and the charge point, a communication mechanism between the CP and a central management system, and finally the creation of a communication channel between the charging system and the energy suppliers (DSO, TSO, smart girds, etc.) are some examples of the communication that takes place at a smart charging architecture [6, 7].

Anomaly Detection

Anomaly detection, which has applications in many different study domains and application areas, is the process of identifying patterns in a set of data that don't match the expected behavior of a system. To defend against threats and attacks on the smart-charging ecosystem, various anomaly detection algorithms are put into use. One type of anomaly detection is the identification of high pricing. This technique can be used to spot a range of threats, including healthcare systems, credit card and general security scams, as well as military systems for spotting hostile behavior. Any extreme values that are observed on the CPs of EV charging during the operation of the CARAMEL's defense mechanism are identified by anomaly detection.

Any deployed IDS must consider collective and contextual irregularities in addition to point anomalies and independent extreme values in the dataset because a system may encounter a variety of anomalies. The latter require a system intelligence to understand the greater environment, whereas the former are oddities that coexist in vast groups.

Time series offer a data overview across time while anomalies can be seen in their graphical representation, which is why they are frequently employed for the deployment of anomaly detection techniques. In order to perform this detection, observation is made for an anomaly or spike that stands out from the rest of the data behavior, such as an upward spike or a sharp draft in the figure. The items of interest, are frequently not unusual things, but rather unexpected bursts of activity like many synchronized charging sessions, especially in the context of exploitation in the smart charging system.

The suggested approach makes use of the flexibility that semi-supervised anomaly detection offers in anomaly detection to datasets that contain both normal and abnormal data without being explicitly labelled. Due to the lack of labels in the dataset, a devised semi-supervised machine learning technique was used to identify anomalous data and build a model to classify any new incoming data. This approach has the benefit of preventing bias against a certain class, which typically arises when datasets are imbalanced (containing more normal than abnormal data) as a result of small sample sizes.

Additionally, because it necessitates simulating assaults on a sandboxing environment to capture both normal and abnormal data, building and annotating a dataset for supervised machine learning algorithms is challenging and time-consuming. Furthermore, trained models with well-known patterns for anomaly identification may not succeed in specific circumstances due to the peculiarities of the training phase and the data that was used for that training. However, unsupervised training is renowned for having low detection rates and a high proportion of false positives (mislabel normal data as abnormal) despite the fact that it does not require labelled data. To cut down on processing expenses and locate every conceivable anomaly, semi-supervised anomaly detection employs dimensionality reduction and other techniques. This combines the best of both worlds [8].

In order to introduce a data-driven strategy for secure smart charging, a number of various techniques are applied to the gathered data. These techniques aim to not only deliver a proof-of-concept demonstration but also to provide some benchmarks for the effectiveness of various algorithms in the field of EV cybersecurity. The incorporation of AI/ML into cybersecurity is essential for the handling of cyber-crisis at the EU level because it gives the opportunity to derive practical insights from massive amounts of data (big data), which would otherwise be impossible for humans to examine [9]. The three different techniques that are used in the scenario of smart charging abuse are:

- **Z-Score (Standard Deviation):** For the different features (e.g., Volume, Duration, ReadingContext) of the dataset that was tested, the investigation of a registration was made (e.g., charging action) and the check if it is within three standard deviations. If not then it is considered as an outlier. An illustration of a normal distribution and its relationship to standard deviation is shown in Figure 2. If connected automobiles are more than 2 standard deviations away from the mean value on a certain day and time, it is a sign of abnormal behavior. For instance, the number of cars on a working day for a specific date and time can be predicted based on the mean value of prior weeks. Although the standard deviation score is not a machine learning technique, it is frequently used as a benchmark for unsupervised ML algorithms

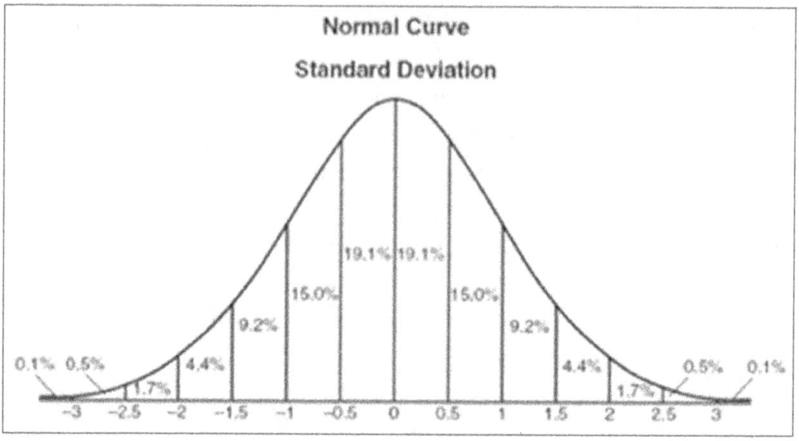

Figure 2. An example of standard deviation applied on normal distribution.

since it is a successful strategy that is simple to comprehend from a human perspective. The method's limitation is that it can only be used with parametric distributions in a low dimensional feature space which may prevent it from being recommended in the event that the feature data gathered by the organization are expanded in the future.

- **Density-based functions:** The idea behind density-based algorithms is that the density of points surrounding an outlier point is much lower than the density of points around an inlier point. The justification is based on the idea that routine activity that can point to a cyber-threat is different from similar events, such as an EV charging on a specific CP, which have (very) similar features. When mapped into a 2-D representation, each point or event has a precise distance from every other point or event, as shown in Figure 3. This distance allows for the creation of several clusters of related occurrences, which may then be used to identify outliers.

  When used on single-dimensional data, the density-based methods also provide an understandable graphical representation (power requested, time of charging, etc.). Although the algorithms are capable of handling multidimensional data and are recommended for multidimensional feature spaces ($n > 3$), the display of the results is difficult for humans to understand. The density-based strategy can have an even bigger impact in future dataset expansions that contain more features. The arbitrary definition of the k variable in this method is a flaw that could lead to issues with overfitting.
- **Isolation Forest:** The ensemble decision tree family of unsupervised learning algorithms includes Isolation Forest, which isolates anomalies rather than grouping related events as do density-based algorithms. The idea behind the approach is that it is possible to separate anomalies since they have extreme

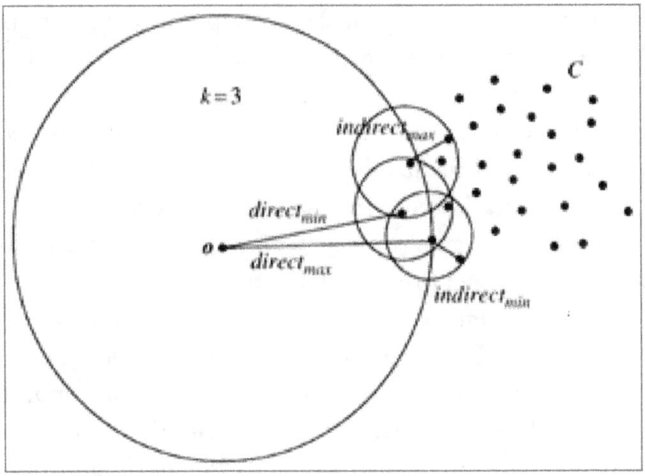

Figure 3. A graphical representation of density-based functions.

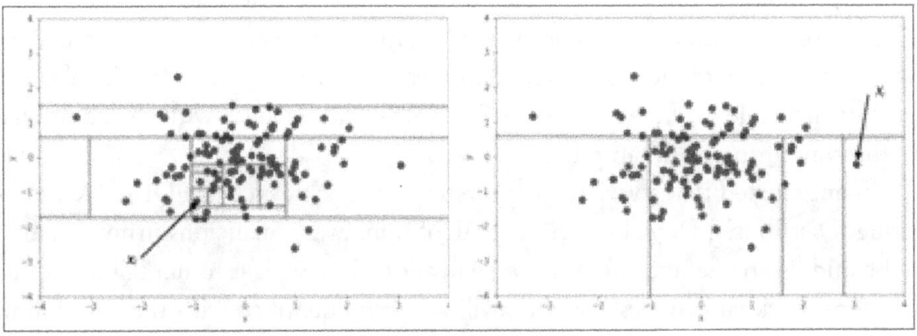

Figure 4. Overview of Isolation Forest Algorithm output.

values in comparison to typical situations. The isolation algorithm is built on iterations that create data sample divisions by first choosing a split value for an attribute, which is then chosen at random. The length of the path is the number of iterations necessary to isolate the path, and this recursive split of the sample can be represented by a tree structure called an Isolation Tree. Figure 4 shows a graphical representation of the algorithm. In contrast to the right side, where the algorithm isolates the outlier point in five iterations, the left side shows the several iterations of the techniques used to isolate an inlier. A charging action that demands an uncommon volume, for example, can be isolated from the algorithm more quickly than a charging action that wants a regular volume since it has unique properties that make it easy to lead to isolation, like the left part of the image.

Because the method considers the high values of an outlier x instance as another data point, the isolation forest technique can perform well even if the training set does not contain any anomalous points. As a result, creating data that simulate aberrant behavior and integrating them into an already-existing database is not necessary.

FDIA, AMI tampering, and data manipulation are typical cybersecurity attack scenarios for PEVs that put the stability and integrity of the electrical distribution network at risk. These attack scenarios could result in inaccurate payment transactions, altered smart meter readings, severe harm to charging stations and automobiles, and the exposure of user privacy information.

Through deep packet inspection, stateful protocol analysis models, and statistical analysis, CARAMEL developed and demonstrated detection strategies for top-ranking and zero-day threats for PEV charging. The first two of them involve processes that assess observed behavior against established profiles of non-harmful protocol activity for each protocol state in comparison to abnormalities discovered through IDS. Establishing stochastic thresholds for diverse kinds of network activity is a prerequisite for statistical analysis. Some of the crucial testing parameters go above predetermined thresholds when an intrusion occurs. It is anticipated that efficient definition of the best sampling intervals for alarm production will take into account the trade-off between accuracy and on-time response with the additional overhead resulting from sampling.

It is anticipated that a variety of cybersecurity attacks, such as FDIA, AMI tampering, Distributed Denial of Service (DDoS), malware, malicious firmware, man in the middle attacks, etc., as well as system anomalies, such as router failures, routing loops, backhaul congestion, etc., will be easily identified with the addition of these modules to a smart charging platform.

Solution Design

The third pillar of CARAMEL focuses on creating cutting-edge AI defenses to safeguard EV charging stations from charging abuse scenarios. The pillar's goal was to develop and put into action a cyber-security framework that can identify any anomalies in the EV system and notify the relevant stakeholders in almost real-time. The pre-trained ML algorithms automatically detect abnormal traffic and convey the alarm across the necessary communication channels, while the visualization graphs give the security administrator an overview of the system's state.

In this use case, the aim was to demonstrate a machine learning pipeline that is capable of detecting anomalies in communication between an EV charge station and its remote back office. From the charge stations that are currently installed worldwide, a part consists of outdated hardware. These legacy charge stations can no longer always be updated (hardware or software) so that they meet the latest security standards. It is, therefore, possible for an attacker to either locally take over the charge station or to intercept the communication in one way or another and thereby, for example, be able to send false smart-charging control signals to such a charge station.

Smart Charging Abuse

Two different forms of attacks are covered under the smart charging abuse scenario. The first one refers to any rising demand that could result in service interruption, and the second type of attack expresses more sophisticated methods through the modification of particular features that could perplex the Intrusion Detection and Prevention System (IDPS) used by EV charging providers to combat hacking. DSOs and TSOs in Europe could experience a series of issues and breakdowns as a result of a potential disruption to GreenFlux's (GFX) service.

The smart charging solution that the CARAMEL solution was tested was provided by GreenFlux and it can secure not only the company's grid but also potentially catastrophic exploitation of the EU's electrical grid by incorporating AI/ML techniques. In the scenario of smart charging abuse, several users synchronize (either purposefully or unwittingly) and move forward with connection/disconnection activities on time, putting an unexpected burden on the electrical grid. The AI/ML methods that were incorporated into the GreenFlux's software, can detect and mitigation actions can be performed to avoid such types of attacks.

A starting point for understanding the behavior of the charging stations, was by extracting the characteristics of a "typical" charging operation, and recognizing suspicious acts as outliers is the data that has been gathered in GreenFlux's lab since 2012. The outlier in this scenario is a charging procedure that cannot be categorized into what is referred to as expected behavior (inlier), such as when an EV asks more power or when a specific car is charged at a different station than normal, displaying odd behavior that has to be further studied.

Either unsupervised machine learning (ML) methods or a set of logic rules that should cover the complete range of what is considered "typical behavior" are used to identify outliers. The available data was used in the framework of CARAMEL, and a number of unsupervised ML algorithms were applied in an effort to identify acts that could pose a threat to the electrical grid. There is no need to go into replicating certain cyber-attack attacks and then use supervised classification algorithms as long as the goal is to identify aberrant behavior that could harm the electric grid. The final annotation of the data for each year was achieved by the application of supervised and unsupervised machine learning techniques. On the basis of the statistics, it is assumed that there will be more false positives. We make this choice to tolerate more false positives in order to collect as many anomalies as possible.

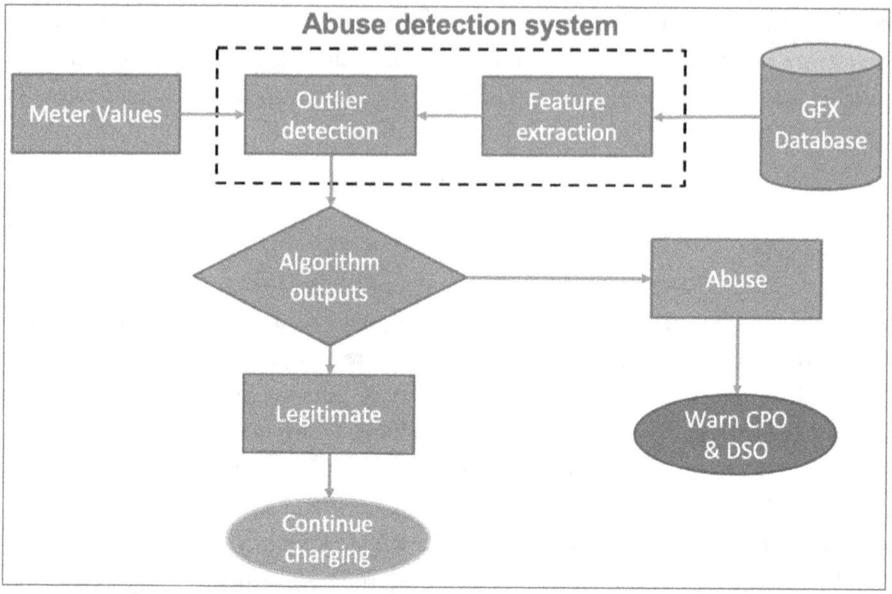

Figure 5. An overview of the Abuse Detection workflow.

In order to find every anomaly in the dataset, we are therefore willing to accept normal points classified as anomalous due to a faulty prediction but not the other way around.

Figure 5 shows the abuse detection pathway, which provides a more understandable method for using ML algorithms to increase the cybersecurity of the EV. The application determines if an incoming event is authentic or suggests a cyberthreat by receiving both real-time data (meter values) and history values from the GreenFlux's database. The Charge Point Operator (CPO) and the DSO are alerted in the latter scenario; otherwise, charging will proceed.

Through the use of anomaly detection algorithms, the AI-enabled cyber-security framework may identify any cyber-security concerns and send a warning to the appropriate stakeholders. Through the use of the clustering algorithms and the visualization of the charging attempts, it also provides a summary of the state of the system. The approach that was followed for the ML pipeline is depicted in Figure 6.

The flow of activities that were followed for the successful demonstration are:

1. Data was collected in the days prior to the attack (were used as regular data for reference).
2. ML component was trained using the aforementioned data.
3. Attack was held against the charging station the day of the demonstration.
4. ML tool successfully detected the attack and the security team was able to address the problem and mitigate the attack.

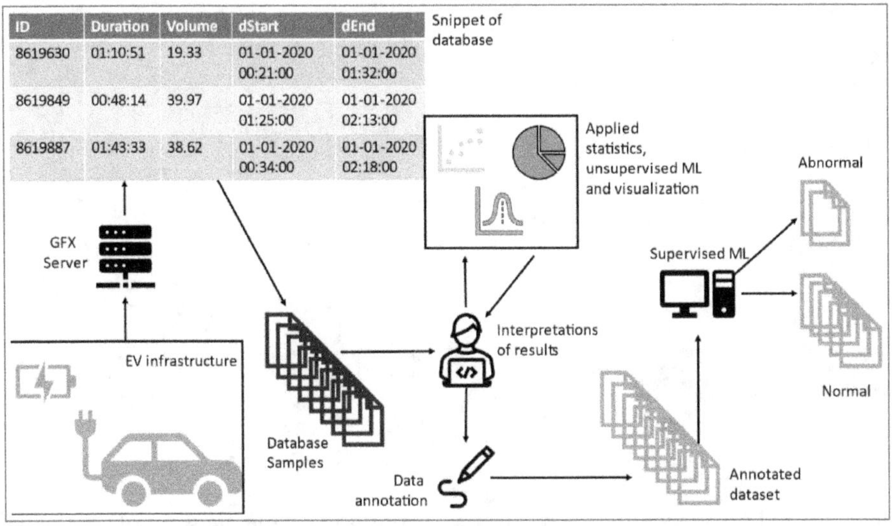

Figure 6. Semi-supervised Machine-Learning pipeline.

Integration and Deployment

The end-to-end communication of all the parties involved in the EV charging process, from the EV charging stations to the energy producers and the EV drivers, is necessary for the integration of the AI-enabled EV cybersecurity framework. The application of various components of the AI-enabled EV cybersecurity framework into current, established GFX processes is a part of the end-to-end communication.

An overview of the cyber-security architecture that is currently set up in the GFX's infrastructure can be found in Figure 7. The smart controller, which is connected to the first protection layer and enables the installed firewall, intrusion detection system, and anomaly detection mechanism, is used to connect the EV charging stations to the electrical grid. The multi-purpose GFX back-office is in charge of managing all incoming data and dispersing it to the appropriate channels. It contains specifics on the energy management strategy, records of the transactions, and information on the smart charging procedures. A complete cybersecurity framework, including a SIEM system and monitoring capabilities, is provided by the CARAMEL analytics, which are plugged on top of the existing system. The high-level architecture of CARAMEL's EV cybersecurity framework gives a general overview of the implemented system, but it omits crucial technical information about the precise locations of the servers, the near-real-time data flow between the various entities, and the implementation of the anomaly detection algorithm.

There is no need for a simulation because the smart charging abuse scenario is based on actual data that GreenFlux has been collecting at its lab in Amsterdam

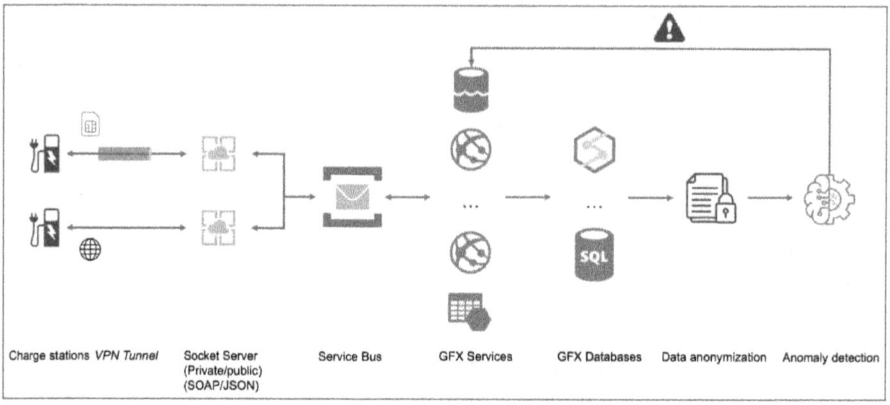

Figure 7. Overview of the demonstration pipeline.

since 2012. The purpose of clustering algorithms is to define divisions of data based on some assumptions, such that the points of a cluster are similar to some ground truth set of classes or the point satisfies the hypothesis that the points present a higher similarity when compared to points outside of the cluster. The combination of supervised and unsupervised machine learning algorithms is the strategy was used. It is essential that a sizable volume of data be annotated, either by skilled human annotators who can spot a system outlier or by simulating anomalous activity on the charging station.

Three separate evaluation strategies were used for the smart charging abuse scenario to guarantee the accuracy of the ML techniques' conclusions. A validation framework that is broad enough to address various facets of cybersecurity in the context of EV smart charging was provided by the applied evaluation methodologies, which combined qualitative (visual inspection, manual inquiry) and quantitative metrics (accuracy, precision).

Setting up a live connection allowed the necessary data to be supplied directly into the anomaly detection tool, which is the major goal of this pilot. Use of the current GreenFlux APIs is not possible due to GDPR constraints (these may only be used by customers that are the owner of the data). Consequently, a pipeline was developed in which the data is first made anonymous before being shared outside.

The following elements make up the communication pathway between the EV cybersecurity framework and charging stations for electric vehicles as shown in the figure above:

- **Charge stations:** They are in charge of keeping track of transactions and measuring energy use. The measurements are delivered to a private Access Point Name (APN) at the telecom operator via the Open Charge Point Protocol (OCPP) protocol over a 2G, 3G, or 4G connection.

- **VPN tunnel from the telecom provider to the WebSocketServer subnet in the GreenFlux Cloud:** This tunnel connects charging stations to the Socket Server Subnet. The VPN tunnel is segregated from the rest of the internet and is encrypted.
- **Service Bus:** The service bus converts incoming OCPP messages into GFX's internal language so that the various services can understand it (and vice versa).
- **GFX Application Services:** The GFX platform is made up of a number of services, each serving a specific purpose. For processing meter data, authorizations, smart charging, etc., there are services available. These services both send messages to the charging stations and process incoming communications.
- **GFX Databases:** Several services use one of the databases to write data to. This is for two reasons. Firstly, for analysis made by GFX and secondly for the fact that this is legally required for particular data (transaction data).
- **Data anonymization:** Before the incoming data from the charge stations can be shared outside of GreenFlux, it must be anonymized after being processed and stored.
- **Anomaly detection tool:** Sending the anonymized data to the anomaly detection tool completes the pipeline.

The communication path between the many elements that interact throughout the EV charging process is shown in Figure 7. After processing some of the data, each sub-component sends its findings to the following pipeline stage.

Transactions Service

Whatever deployment is employed, the messages follow the same path. The "Transactions Service," which is in charge of handling Start and StopTransaction messages (which form the core of the CDR dataset), as well as the MeterValues (MV) Service, is the most intriguing service (the basis for MV dataset). The Transactions monitors the socket server for new messages. The way an incoming communication is handled varies depending on its nature.

Start Transaction Event

StartTransaction messages from Charge Stations are processed by this function after being received via the Socket Server. It will obtain charge station data from the relevant domain, confirm transaction authenticity using the Token Service, and produce a transaction document.

Stop Transaction Event

The Socket Server sends StopTransaction notifications to Charge Stations, which this function processes. The charging station will be obtained from the CPO domain, the transaction will be authenticated against the Token Service, and the transaction will be finished.

MeterValue Event

The MeterValue messages from Charge Stations that are received via the Socket Server are processed by this function. The charge station will be obtained from the CPO domain, the meter values processed, the charging period updated, and the MeterValue stored in the relevant database.

Details on the Dockerization Process

The technical information on how the EV cybersecurity framework was implemented utilizing Docker technology to offer a service that hosts various software components in packages using virtualization is described below. Each software component is referred to as a "container," which runs independently from the other containers but has the ability to request to perform particular activities. Six distinct containers will be created for the EV cybersecurity framework, each serving a particular function in the communication chain.

The fundamental design of Docker is shown in Figure 8. The many created software sub-components are displayed at a higher level. Every component can be run at various intervals, giving security administrators the flexibility to monitor the system's operation. The pseudo-generated traffic gives a simulation of active EV charging processes that match the pattern of genuine historic traffic, which is more unique to the nature of the software sub-components. When receiving batches of the fictitious traffic, the data pre-processing function is enabled at regular intervals. The third software sub-component uses unsupervised machine learning (ML) methods to cluster the generated traffic before sending the results to the next sub-components, which are displayed, in order for the security administrator to see an optical depiction of the system. When instances of the incoming traffic exhibit the same pattern as the anomaly-annotated examples in the training dataset, the supervised ML algorithms label those occurrences as anomalies. The container engine, which operates the container and receives pre-defined requests including command-line arguments, makes up the second layer of the architecture. Finally, the infrastructure layer comprises the databases, the processing units, the memory capacities, and the network connections. The operating system layer enables the execution of many containers in the same architecture without interfering with them.

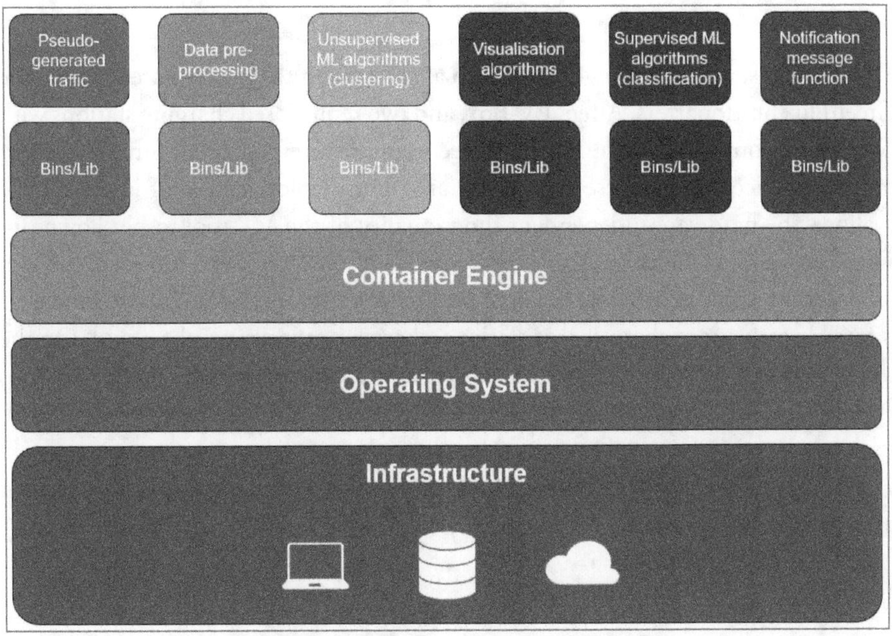

Figure 8. The docker architecture.

Experimental Setup

The communication chain between the various subcomponents was briefly discussed in the preceding sections, along with the integration and deployment of the cybersecurity framework into the GFX infrastructure. The hardware and software components that were used for the demonstration and evaluation of the EV cybersecurity framework and were employed for the CARAMEL project are presented in greater technical depth in this paragraph.

The deployment stage is divided into three parallel parts, specifically:

1. Charge stations at the GFX Office parking are functional charging stations that may be used remotely to mimic various danger scenarios.
2. The GFX test facility's charging stations offer a variety of current and vintage EV charging station models as well as an EV test simulator that can simulate various charging patterns.
3. To replicate all connection between the GFX platform and a charging station, virtual charge stations were used.

The many experiment configurations provided more flexibility for the scenario's actual demonstration. In the following sections, a more thorough explanation of the various experimental setups is described.

Charge Stations at GFX Office Parking

Six charging stations with two plugs each are located in the parking area next to the GreenFlux headquarters. Alfen, EV-Box, and two repurposed charging stations with GreenFlux controllers make up the three manufacturers in total. Simple assaults can be practiced using these charge stations. This deployment is of interest since it allows the hardware-independent functionality of the ML tool to be shown off. These charging stations are also actively used by staff members of GreenFlux and other local businesses, therefore actual production data is produced here. Finally, it is feasible to manage these charge stations' charging capacity via the cloud. On the cloud platform or smart charging signals, an attack can be simulated in this fashion. Simply said, it is impossible to launch a hardware attack by prying open the charge stations.

Figure 9 shows an EV charging station that is always in use at the GFX office lot in Amsterdam.

Figure 9. EV charging station at GreenFlux parking area.

Test Facility

Several charging stations are often employed for testing in the GFX test facility. There are about ten different charge stations available overall from various manufacturers. We provide both current and vintage models (legacy means: charge stations that are still in the field but have outdated hardware that cannot be upgraded to the latest security standards). With this, various scenarios can be created, each of which causes a distinct kind of anomaly in the final dataset.

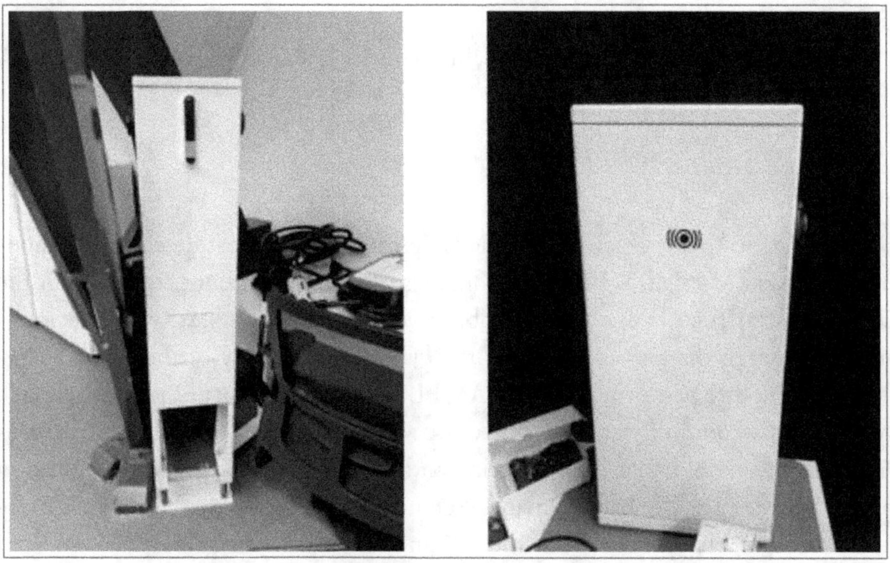

Figure 10. Charge stations in the GFX test facility.

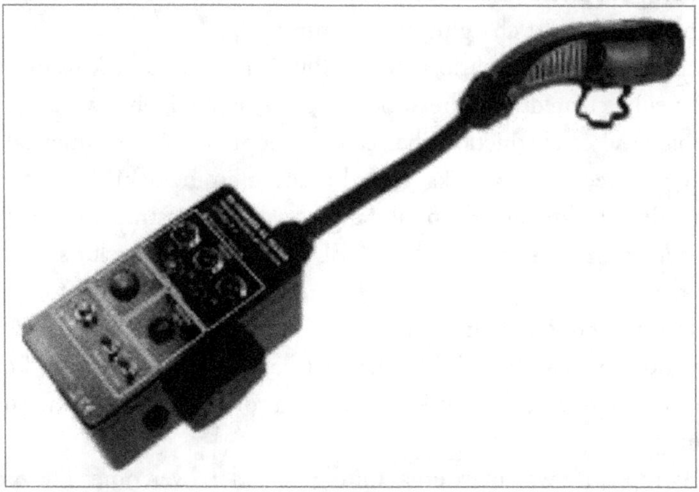

Figure 11. EV charging simulator with schuko outlet.

The facility features an EV test simulator that can communicate with charging stations just like a genuine EV would. In order to mimic more complex attacks, it can modify the analog signal between the charge station and the EV. The charge station may transmit so-called StatusNotifications if, for instance, there are too numerous or irregular state transitions within a short period of time.

Figure 10 presents two charging stations in the GFX test facility.

The simulator also has a schuko contact that may be used to connect any kind of load. With the help of this, it is possible to manipulate the energy meters in the charging station and produce odd MeterValues (that were used for the datasets to train the ML tool).

An EV charging simulator with a schuko outlet is shown in Figure 11.

Conclusions and Future Work

For plug-in electrical vehicles, CARAMEL created cyber threat detection methods in the context of Pillar 3. Finding anomalies in communications to and from the charging infrastructure was the main objective. Due to its complexity and the interactions between the parties involved, the cybersecurity study of the smart charging infrastructure performed by CARAMEL revealed a number of potential attack vectors. This communication architecture was vulnerable to a number of security risks on several levels. The metering and payment for energy, communication breaks between the EV battery management system and the charge point, a communication mechanism between the CP and a central management system, and finally the establishment of a communication channel between the charging system and the energy suppliers (DSO, TSO) are some examples of the communications that take place at a smart charging architecture.

The objective of this scenario is to rapidly identify an attack using a machine learning pipeline created to address the use case's requirements while also offering a respectable level of abstraction that can be adopted from commercial partners. The ability to detect these attacks, record them in an incident database, and alert the system administrator to potential dangers was demonstrated in the GFX software with the integration of the CARAMEL solution which includes the ML algorithm for the Smart Charging Abuse. The technical information for putting the EV cybersecurity framework into practice was offered in the previous subsections. This subsection discusses the difficulties that have been overcome and offers suggestions for future research that might result in a comprehensive framework for EV cyber-security.

GDPR compliance was an external hurdle we had to overcome. The most recent rule stipulates that pseudo-anonymization of the data must occur before sharing it with any third-party organizations. To get around this restriction, the pseudo-anonymization process must either happen in real-time in the GFX infrastructure or simulated traffic needs to be created. We can use the pseudo-generated software component to deploy the CARAMEL EV cybersecurity framework everywhere, subject only to applicable laws.

The physical interaction with EV charging stations is the second major barrier. The suggested framework must be implemented in production units for all of the available charging stations in order to be fully end-to-end deployed. Multiple actors must coordinate their actions, and scaling problems arise during the design and testing phases. We can free up human resources and conduct experiments in more manageable case study scenarios by using simulated or hybrid environments.

The scalability difficulties will be resolved in a future development of the framework, enabling interconnection between the EV charging stations and the various players involved in the charging process, from EV drivers to energy producers. Future EV frameworks that enable a faster smart charging procedure could be built on the suggested framework as a starting point.

Attacks against smart charging can affect any part of the energy ecosystem directly or indirectly (charging infrastructure, TSO, DSO, etc.). The absence of security features in the charging stations for identifying and preventing potential assaults and threats can undermine the dependability and security of the entire energy supply network. For instance, the DSO might be impacted by a hypothetical attack vector in the smart charging system, which would cause a partial energy blackout. The interconnectedness of many players utilizing various technologies creates access to dangers and vulnerabilities as CPOs and e-Mobility Service Providers (eMSP) struggle to integrate Artificial Intelligence approaches to modernize their services.

A Machine Learning ML pipeline has been created within CARAMEL to recognize aberrant behavior throughout the charging process using a real dataset of a typical EV charging firm. Software developers, security administrators, and electrical engineers can be made aware of potential dangers to the smart charging infrastructure by this proposed security pipeline. The relationship between various actors in real-time alerting for breaches, known vulnerabilities, and zero-day assaults is a topic that should be investigated in the future. The development of a shared repository that stores, updates, and ranks known threats and attacks is an extension of this information exchange. The OCPP packets could be used as part of the planned ML pipeline to build a library of signature-based attacks inside the private Azure vNET where the GFX socket server is located that are capable of thwarting any malicious packets. Additional headers could be utilized to provide a more thorough perspective of the incoming transactions through the inspection of the OCPP packets.

References

[1] IEA (2022), Global EV Outlook 2022, IEA, Paris https://www.iea.org/reports/global-ev-outlook-2022, License: CC BY 4.0

[2] IEA, *Slow publicly available chargers, 2015-2021*, IEA, Paris https://www.iea.org/data-and-statistics/charts/slow-publicly-available-chargers-2015-2021, IEA. License: CC BY 4.0

[3] Chaudhry H., Bohn T., Security concerns of a plug-in vehicle. In: Innovative smart grid technologies (ISGT), 2012 IEEE PES, IEEE, pp 1–6, 2012.

[4] Hoppe T., Dittman J., Sniffing/replay attacks on CAN buses: a simulated attack on the electric window lift classified using an adapted CERT taxonomy. In: Proceedings of the 2nd workshop on embedded systems security (WESS), 2008.

[5] Kleberger P., Olovsson T., Jonsson E., Security aspects of the in-vehicle network in the connected car. Intelligent vehicles symposium (IV), 2011, IEEE, pp. 528–533, 2011.

[6] Clairand, R.-G. &. (2018). Smart Charging for Electric Vehicle Aggregators considering Users' Preferences. *IEEE Access*, 1–1.

[7] Wen, Y. G. (2017). Fog Orchestration for Internet of Things Services. *IEEE Internet Computing*, 16–24.

[8] Xue, S. &. (2010). Semi-supervied outlier detection based on fuzzy rough C-means clustering. *Mathematics and Computerrs in Simulation, 80*(9), 1911–1921.

[9] Artificial Intelligence – An opportunity for the EU cyber-crisis management, Workshop, ENISA, June 2019, https://www.enisa.europa.eu/events/artificial-intelligence-an-opportunity-for-the-eu-cyber-crisis-management

DOI: 10.1561/9781638280613.ch4

Section 4

Remote Control Vehicle

Introduction

Remote driving or so called tele operated driving of a highly automated vehicle (i.e. SAE L4 automated vehicle according to SAE J3016 levels of driving automation) occurs when the connected automated vehicle (CAM) is remotely controlled by a remote human operator when the CAM detects a failure in a critical subsystem or faces out of operation design domain such as sudden weather change, non-signalized intersection, or work-zone [3GPP TR22.886]. A number of use case scenarios related with road safety, business is available utilizing remote driving of CAM including[1,2]:

- Re-starting after stopping on the shoulder at a station. The remote driver checks that there are no vehicles around, especially from behind, and operates the "start" operation,

1. 5G MOBIX Deliverable D2.1, "5G-enabled CCAM use cases specifications", 2020.
2. 5G Americas white paper, "Cellular V2X Communications Towards 5G", March 2018.

- Providing remote support service to CAM when out of ODD due to sudden weather change such as heavy rain, fog or other environmental conditions (i.e. work zone),
- Joining the priority road from a side road, no traffic lights. The vehicle automatically stops at the stop line, the remote driver checks the safety and "starts" operation,
- Facilitating tele-operated driven low speed autonomous shuttles or ballet parking services with predefined routes and stops;
- Providing remote driving services for individuals who are unable or unlicensed to drive (e.g. youth, elderly, disabled persons etc.);
- Providing a fall-back driving solution for autonomous vehicles which have encountered unfamiliar navigation environments or developed some faults.

In a typical remote driving use case, the remote human operators refer to the vehicle status (i. e. speed, RPM, gear position, and etc.) and video sensing information around the vehicle. And send back commands for controlling the vehicle in a more reliable manner over a V2N connection between an L4 vehicle and a Remote Operations Centre. The reliability and effectiveness of the commands from a remote human operator of an L4 vehicle is contingent on the quality and timeliness of the data received from the vehicle's sensor feeds. Therefore, any significant constraints or disruptions in the sensor data transfer would not be tolerable in a remote driving scenario. For instance, limits on the uplink throughput in a V2N connection would limit sensor data feeds in terms of achievable resolution, frame (or refresh) rates and compression applied (contributing compression latency). These throughput limits become more severe in road environments with multiple autonomous vehicles contenting the same mobile network resources for V2N communications. The constraints in achievable uplink throughput and area capacity in legacy 4G networks would make wide-scale adoption of the solution challenging.

The increased availability of 5G connectivity will alleviate the aforementioned throughput constraints and reduce latency, as well as, providing additional benefits of security, service discovery and so on. These improvements are an enabler higher-resolution sensor data feeds (and possibly more feeds) from vehicle to human operator in Remote Operations Centre.

The remote driving of an L4 vehicle is enabled by a V2N connection between the vehicular Onboard Unit (OBU) and a remote server hosting V2N applications, in this case the remote driving application used by the remote human operator. The V2N connection transfers the sensor data feed (high resolution perception data) from the vehicle to the remote human operator (in the uplink direction). The sensor data provides the human operator a "driver's view" for that particular vehicle and allows the human operator to send appropriate command messages (e.g. command trajectories) back to the L4 vehicle (in the downlink direction).

The remote control/driving of vehicle presents stringent requirements on connection between the vehicle and the Command and Control Center. These requirements include the need to ensure that the human operator always maintains connectivity to the vehicle they control, the latency minimized to ensure timeliness of the downlink control messages from the human operator; and the uplink capacity is guaranteed for the transmission of the sensor data feeds from the vehicle. The whole control loop needs to be kept tight. The accumulated delay from: sensor reading, sensor data processing, uplink, data visualization, manual control, control signal reading, downlink, and control signal processing to control has to be kept low for direct control (depending on speed and dynamics of the vehicle). Furthermore, the vehicle should be aware of any latency issues, so that the operational speed could be adjusted accordingly.

Remote driving (and other V2X use cases) will occur in multi-operator scenarios in legacy 4G [3GPP TR 36.885][3] and future 5G [3GPP 38.885][4] contexts, whereby the L4 vehicle trajectory is an area covered by multiple public land mobile networks (PLMNs) or transitions between two PLMN coverage areas. Therefore, the remote driving use case considers two possible scenarios depending on the PLMN used for realizing the V2N connection.

In pillar 4, Korean partners have engaged in research and development of cyber-attack detection for remote control vehicle (RCV) environment. The general solution of security authentication and authorization can also be applied in RCV environment and detailed research efforts are covered by the previous pillars. Thus, we will not repeat similar efforts for pillar 4 but, instead, will focus on RCV specific solution. RCV has its unique characteristics which need to identify its specific requirements. We conducted our efforts to identify such requirements, defined a functional architecture based on these requirements, machine learning based detection mechanisms and followed by experiments with verification. We described our resulting efforts in the following sub-sections.

State of Art/Innovation

While an automated driving vehicle needs a lot of sensors and algorithms such as obstacle detection and avoidance, remote driving with human operator should

3. 3GPP TR 36.885, "Study on LTE-based V2X Services" June, 2016.

4. 3GPP TR 38.885, "NR; Study on Vehicle-to-Everything" March, 2018.

be realized using less of them. A main objective of remote driving is to control or to drive an automated vehicle remotely when the automated vehicle is malfunctioned or the driver in the vehicle is in an accident such as a heart attack. Since the remote driving vehicle is able to share the real time live video stream not only around the vehicle but also inside the vehicle, driver or passengers in the remote-controlled vehicle should be safer during the driving to their final destination.

The table below shows an overview of the use case for remote driving.

Use Case Short Name	Remote Driving
Objective	Providing reliable and safe control of the remote driving by high-throughput data sharing between remote driving vehicle and remote site operator
Actors	Remote Driving Vehicle (RDV), Driver, Remote Operator
Pre-conditions	Vehicle is parked or stopped on the road with stable 5G-connectivity to remote site
Use Case flow	1. The driver in the remote driving vehicle (RDV) requests a remote operation from the remote operator. 2. The remote operator checks the status of the RDV (speed, video, state of fuel, automated vehicle controller, and so on) and decides to activate the remote driving function from the remote site. 3. The remote operator starts to control or to drive the RDV with limited speed. 4. The remote operator drives the RDV to a safe area. 5. When the RDV is parked, the remote operator checks the status of the RDV.
Post conditions	The RDV is parked in a safe area.

Applying machine learning technology into cyber-attack prediction and detection in the general network such as Internet and mobile networks has been widely researched and various solutions were developed. However, R&D of the same topic in the autonomous vehicle environment is relatively new, especially for RCV. We tried to utilize state-of-the-art solutions by adopting and extending into our RCV environment to provide an optimal solution. Our initial efforts to research and develop a solution and its details are described in the following sub-sections.

Threats/Problems Considered/Detected

The threats and anomaly considered in RCV environment has been considered with the following two use cases. The first use case is an attack on RCV control and status message transmission and another use case is attack on RCV status image transmission. The details on two threats use cases are described in the tables below. Among two use cases, we have experimented use case 1 at this stage. The use case 2 will be further studied in the next year.

Scenario Name	**Attack on RCV Control & Status Message Transmission**
Related Pillar	Connected Mobility
Scenario Descriptions	There are two scenarios: (a) Control message from a control center to a RCV is attacked, (b) Status message from a RCV to a control center is attacked.
Challenges	• Deploy a PKI system to register RCVs, distribute security credentials, authorize vehicles to transmit signed messages, revoke certificates and distribute lists of revoked certificates. • Ability to detect fake messages: not signed, signed with a non-valid certificate, signed with revoked certificates, replayed and non-authorized.
Assumptions & Preconditions	• The scenario is provided with a communications infrastructure: namely OBUs, RSUs, Small Cells and RCV control (currently proprietary) communications protocol suite. • RCVs are equipped with an HSM to store cryptographic material and a location system.
Goal (Successful End Conditions)	• RCVs drop fake messages and prevent safety applications from being misinformed. • Control center identify fake status messages and prevent from RCVs being misinformed
Involved Actors	• Malicious attacker • Fixed infrastructure • Outside infrastructure
Scenario Initiation	• The cyber attacker transmits different types of fake control messages to RCV • The cyber attacker transmits different types of fake status messages to Control center

Scenario Name	Attack on RCV Control & Status Message Transmission
Main Flow	Case (a) Fake messages: 1a. Normal RCVs register to the PKI system and transmit and receive messages. 2a. The attacker sends fake messages. 3a. Normal RCVs check the signature of these messages, detect which messages are not compliant and drop them.
Novelty	This scenario will implement a complete security infrastructure for a RCV communications system: PKI servers with capacity to distribute ATs and revoked certificate lists, message signature ability in the OBUs and a machine learning based algorithm to choose when a vehicle has to change the AT to avoid being tracked.
Evaluation Criteria	Case (a) fake messages: Normal RCVs detect all non-compliant messages.

Scenario Name	Attack on RCV Status Image Transmission
Related Pillar	Connected Mobility
Scenario Descriptions	Status images from a RCV to a control center is attacked.
Challenges	• Deploy a PKI system to register RCVs, distribute security credentials, authorize vehicles to transmit signed messages, revoke certificates and distribute lists of revoked certificates. • Ability to detect fake status images: not signed, signed with a non-valid certificate, signed with revoked certificates, replayed and non-authorized.
Assumptions & Preconditions	• The scenario is provided with a communications infrastructure: namely OBUs, RSUs, Small Cells and RCV control (currently proprietary) communications protocol suite. • RCVs are equipped with an HSM to store cryptographic material and a location system.
Goal(Successful End Conditions)	• RCVs drop fake status images and prevent safety applications from being misinformed.
Involved Actors	• Malicious attacker • Fixed infrastructure • Outside infrastructure

Scenario Name	Attack on RCV Status Image Transmission
Scenario Initiation	• The cyber attacker transmits different types of fake status images to Control center
Main Flow	Case (a) Fake status images: 1a. Normal RCVs register to the PKI system and transmit and receive messages. 2a. The attacker sends fake messages. 3a. Normal RCVs check the signature of these messages, detect which messages are not compliant and drop them.
Novelty	This scenario will implement a complete security infrastructure for a RCV communications system: PKI servers with capacity to distribute ATs and revoked certificate lists, message signature ability in the OBUs and a machine learning based algorithm to choose when a vehicle has to change the AT to avoid being tracked.
Evaluation Criteria	Case (a) fake status images: Normal RCVs detect all non-compliant status images.

Solution Design

1. General Architecture of Remote Control Vehicle

Remote driving has to be provided with multiple up-link video streaming with ultra-low latency and down-link control signal with low latency at the same time. Therefore, 5G key technologies such as eMBB (Enhanced Mobile Broadband), URLLC (Ultra-Reliable Low Latency Communication) should be developed. The key requirements of the mmWave communication for the realization of the remote control vehicle use case are downlink and uplink data rates and can be summarized as follows:

- Downlink data rate required to transmit real-time vehicle control information: 1 Mbps
- Uplink data rate required to transmit real-time video and vehicle data: 50–100 Mbps
  - ✓ A total of 8 driving cameras were installed on the front, rear, and side of the vehicle, and at least 4 cameras were used for the demonstration.
  - ✓ According to the 3GPP TR 22.886, the uplink data rate required to deliver two videos with H.265/HEVC HD video codec is 25 Mbps.

Remote driving use case enables remote operators to access the right of control in case of an automated vehicle malfunction or when the driver is in an accident.

The most important factors for realizing remote driving should comprise the following: Ensuring enough field of view and high definition of view for front camera, ultra-low latency to sharing live video stream between vehicle equipped cameras and remote site, and reliable connectivity to control remote driving vehicle in remote site. Consequently, remote driving vehicles need to share not only driving information like speed, position, and videos (front, right and left side, and rear), but also vehicle status information like steering angle, gear position, throttle pedal position, and fuel consumption with remote operators. The driving and status information provided by the remote driving vehicle should be transmitted to the human operator at the remote site with ultra-low latency. In order to share high definition live video stream data with remote sites in real-time, a very high up-link data rate should be required and it will be realized by the 5G network. At the same time, the control data to drive a remote vehicle should be generated by a human operator at the remote site and be streamed to the remote driving vehicle through down-link with low latency. The Figure 1 depicts the overall system architecture of the remote control vehicle.

As a deployment scenario for remote driving use case, a remote driving vehicle has connectivity to the 5G network through a base station (BS) including micro-cell BS, and BS-type road side unit (RSU). In-vehicle UE that is equipped in the remote driving vehicle will provide the connectivity to the 5G network by connecting with RSU.

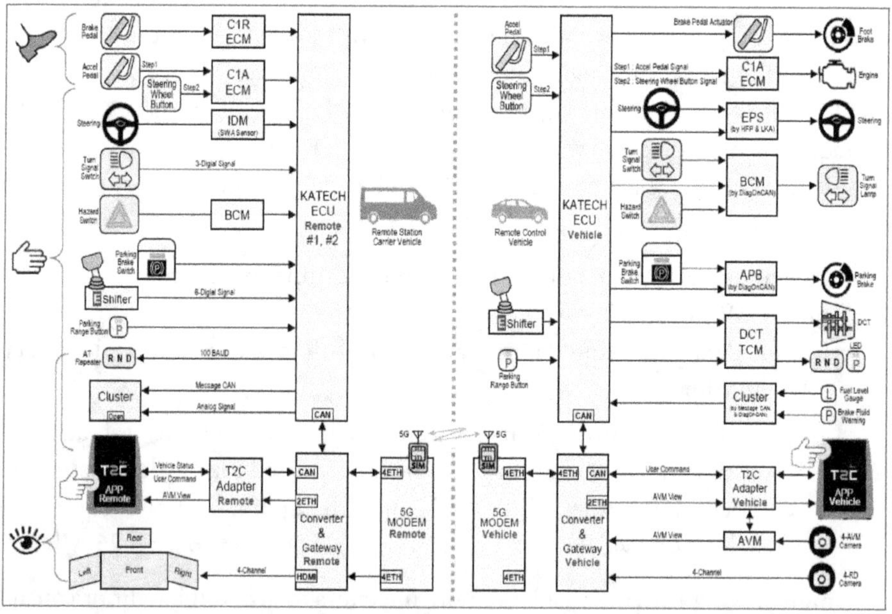

Figure 1. Overall system architecture of remote control vehicle.

The remote driving use case generally supports enhanced mobile broadband (eMBB) and ultra-reliable and low latency communication (URLLC) 5G-service to stream not only vehicle status data and high definition video data from remote driving vehicle to remote control station but also stream RCV control message from remote control station to remote driving vehicle at the same time.

2. RCV Functional Components

Arkana by Renault Korea Motors is used for a remote-controlled vehicle use case. Totally 8 HD cameras (4 sets of cameras for a real-time video stream of the front, left/right, and rear side video, and another 4 sets of cameras for a real-time video stream of around view) and 1 DGPS are mounted in the test vehicle. There are five V2X modules for video data processing (LVDS to Ethernet) and it is connected to the mmWAVE OBU through a network router. HMI (T2C) is connected to the network router for displaying all sensing, vehicle status data and Digital Tago-Graph (TDG) is connected with in-vehicle network to log major vehicle status data including speed, rpm, break signal, aceel pedal signal, steering wheel angel, brake status, gear lever position, operation mode of RCV.

To carry out the trials for the remote control vehicle use case, the automated driving functions have been adapted. The test vehicle shares its each front, left and right, and rear video information and also its surrounding view video information with a remote control station based on mmWave communication. CAN communication is mainly used to aggregate not only location data from DGPS but also vehicle status data (i.e. vehicle speed, steering angle, break) from the in-vehicle network of the remote control vehicle. These CAN data is converted to Ethernet packets by CAN to ethernet converting module while V2X module converts video data(LVDS) to Ethernet packets. All sensor data made with an ethernet packet is aggregated with video data by a network router and transmitted to a remote server via mmWAVE on board unit. The Figures 2 and 4 shows overall architecture and sensor implementation of the RCV.

The mmWave-based OBU, also called User Equipment (UE), is installed in the test vehicle. The control plane (CP) and User plane (UP) protocols have been integrated across L1, L2, and L3 functions as depicted In Figure 4. L1/RF and L2/L3 functionalities of the vehicle UE are performed in separate modules. Similar to the gNodeB, the UE L2/L3 protocol stacks are implemented by software on a 2:2 GHz Intel XEON processor (D-2183IT). Since the same baseband board is shared by both the DU and vehicle UE, the UE L1 functionalities are implemented on the FPGA chips, similar to the DU. The main differences of the vehicle UE from the gNodeB-DU are the RF and antenna parts. The vehicle UE has a beam switching capability. To do so, the vehicle UE has three transmit and receive array antenna pairs, each of which consists of vertically and horizontally polarized components.

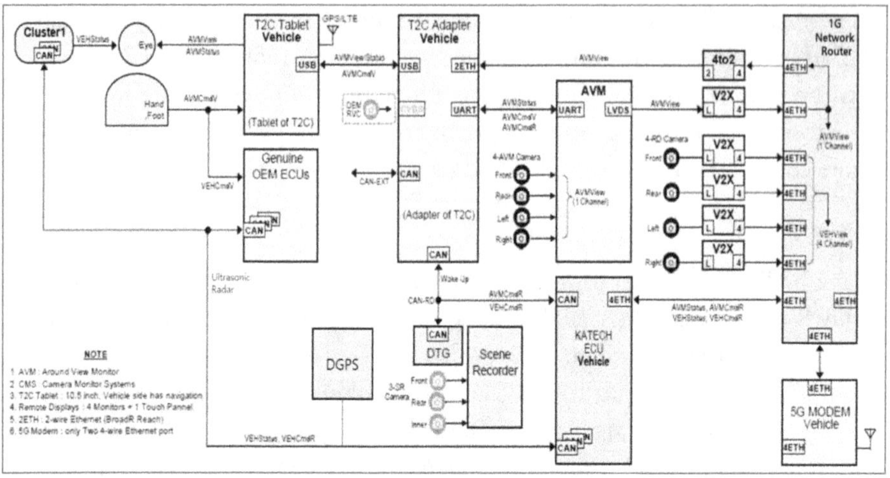

Figure 2. Overall architecture of sensor system of test vehicle.

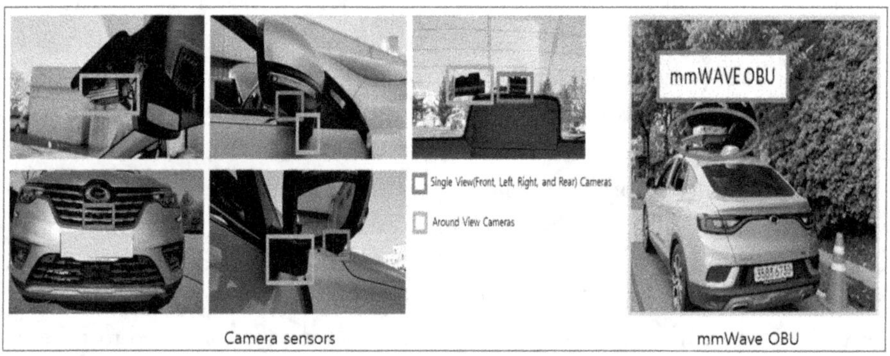

Figure 3. Sensor implementation of the RCV (Arkana/Renault Korea Motors).

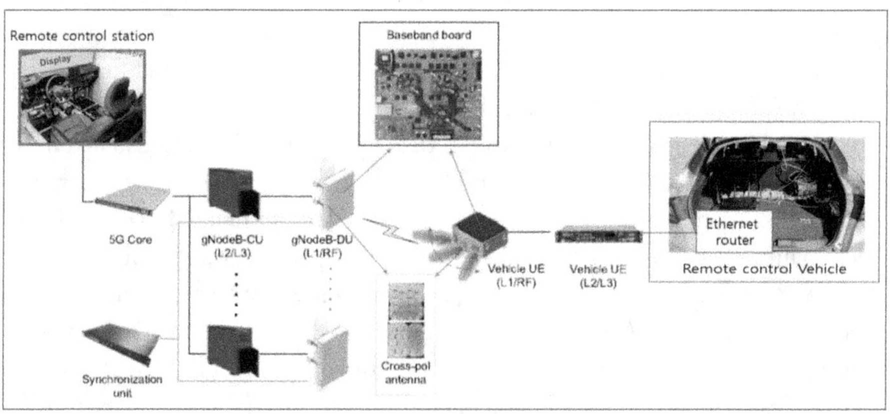

Figure 4. mmWAVE OBU integration.

The beam switching is controlled by an L1 controller based on the RSRP measurements. The beam switching time is measured less than 1 us. In-vehicle UEs (e.g., smartphones) can access the network through a WiFi access point (AP) installed inside the vehicle, which is universally available and has a small coverage within a vehicle. The vehicle UE has a beam switching capability. To do so, the vehicle UE has three transmit and receive array antenna pairs, each of which consists of vertically and horizontally polarized components. The beam switching is controlled by an L1 controller based on the RSRP measurements. The beam switching time is measured less than 1 us.

3. RCV Control Station Functional Components

Since accuracy of generated control input is important for remote control vehicle, real automotive parts related with not only control input such as steering wheel angle, acceleration/brake input, gear position, truing signal but also display unit to display vehicle status data such as digital cluster are used to implement the remote control station as shown in Figure 5.

There are six automotive parts to generate control inputs that are transmitted to RCV as follows;

- Intelligent driving module (IDM): steering wheel angel
- C1R engine control module (ECM): acceleration pedal signal
- C1A engine control module (ECM): brake pedal signal
- Body control module (BCM): hazard lamp signal
- KATECH electric control unit (ECU): turning signal, gear positioning signal, parking brake signal

There are three automotive parts and four external monitor to display status data received from RCV as follows;

- Digital cluster: RCV status data (i.e. Speed, RPM, Gear position, turning signal, hazard lamp, and etc)
- T2C cluster: RCV operation status data (i.e GPS location, communication status, RCV service status, etc)
- Display monitor: HD video data (front, left, right, rear, and around view)

Converter and gateway ECU in the remote control station converts all control input to TCP/IP protocol and transmits it to the remote control vehicle via mmWave vehicular communications.

4. RCV Threat Detection and Analytics Functional Architecture

Figure 6 illustrates the RCV threat detection and analytics functional architecture. It consists of three layers of functional elements: data collection, data pipeline,

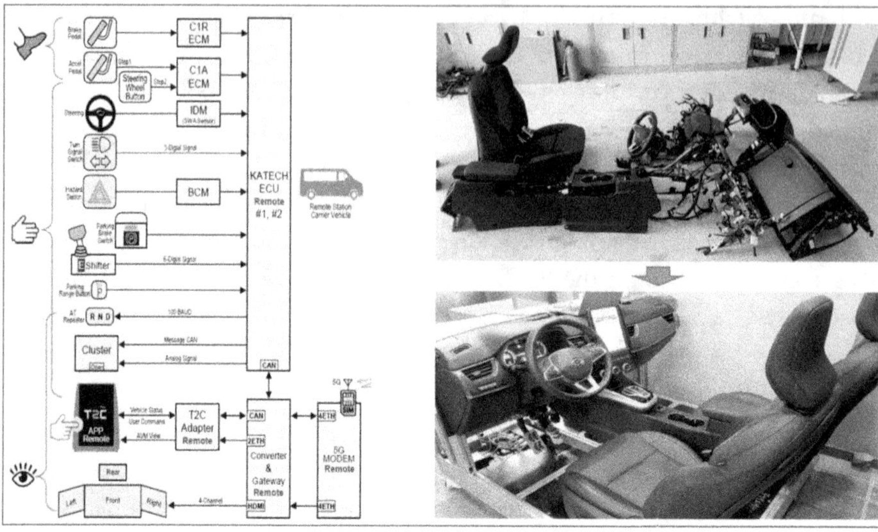

Figure 5. System architecture of the remote control station (left) and implementation of the remote control station (right).

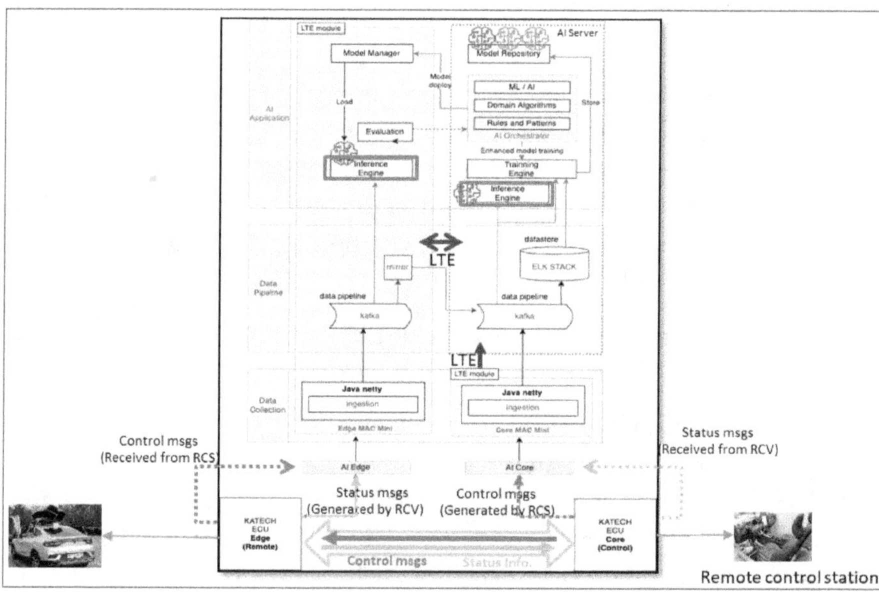

Figure 6. RCV threat detection and prediction functional architecture.

and AI application layers. Data collection layer provides functionality for collecting RCV messages and events exchanged and occurred between RCV and its remote control center. There are two monitoring points one at RCV attachment port to the network and another at the remote RCV control center attachment port to the network. The AI edge functional element is located at the RCV side and the AI core functional element is located at the remote RCV control center side. All messages and events are collected in both directions from the remote RCV control center and RCV and vice versa.

Messages and events at RCV side are collected and parsed at the ingestion module of AI edge data collection layer and sent to data pipeline layer to forward the same copy to remote control center in order to compare them at both ends. The forwarded messages and events are stored in the datastore to be used for ML training. Our solution uses KAFKA to stream the collected messages and events in real-time. Also the data pipeline layer further sends the messages and events to the inference engine in the AI application layer to predict and detect any cyber attacks in real-time. The machine learning model is generated by AI core and deployed in the inference engine of AI edge.

Messages and events at the remote control center are also collected and parsed at the ingestion module of the AI core data collection layer and sent to the data pipeline layer. They are stored in the datastore to be used for ML training. The data pipeline layer also sends them to the inference engine in the AI application layer to predict and detect any cyber attacks in real-time. The same ML model used in AI edge is deployed in the inference engine of AI Core. Additional important functional element in the AI core is the AI orchestrator. It receives various inputs from AI edge and AI core and performs training to generate the optimal ML models. Various ML algorithms are used to train and generate the ML models.

5. RCV Threat Detection and Prediction Mechanisms (TS Choi)

Cyber attacks in the RCV environment can happen at various locations. Our solution assumes the following three scenarios. Cyber attacks can occur at the network in the direction of the control center to RCV. Second case is similar but the direction of the attack is from RCV to the remote control center. Lastly, cyber attack occurs at the RCV internally. Figure 7 illustrates such three scenarios.

Figure 8 describes our cyber attack prediction and detection model. It consists of four steps: lead-lag analysis, data pre-processing, cyber-attack prediction and detection. In the lead-lag analysis step, the relationship between lead and lag of input data collected from various sources is analyzed.

In the lead-lag analysis step, the time difference between when the control message is sent and when the status result is received is analyzed. It is typical that the delay between control and action can happen. Therefore, it is a very important process for the design of prediction models.

In the pre-processing step, conversion of data sets collected into an appropriate form for the model input. For the RCV environment, since control messages arrival and actions executed in the vehicle are not synchronous, it is required to synchronize the two events' timing. Pre-processing takes care of the issues by proper sampling

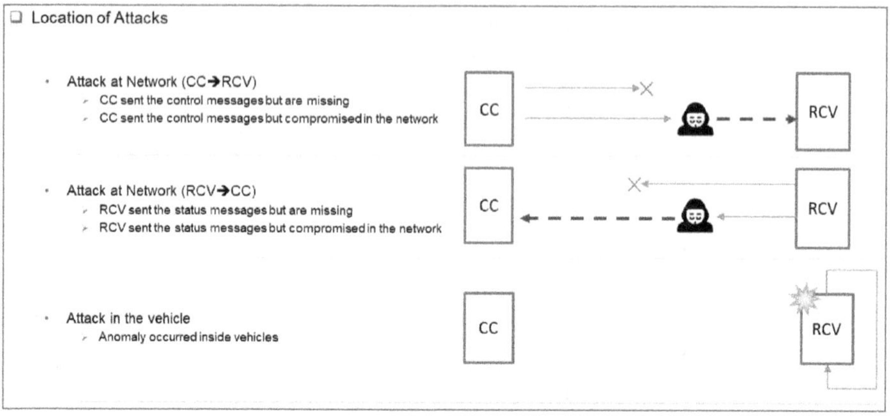

Figure 7. RCV cyber-attack scenarios.

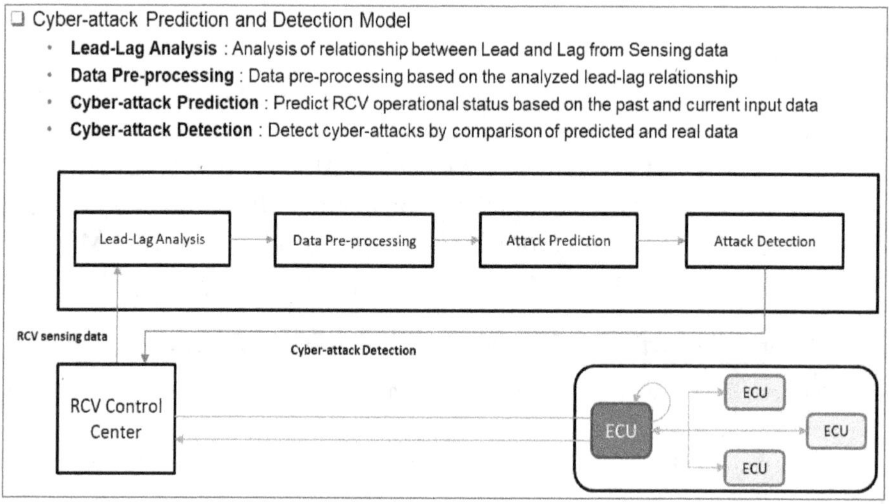

Figure 8. RCV cyber-attack prediction and detection model.

of the data based on the observation. As the next step, the time synchronized data which is measured at a particular point of time needs to be converted into time-series data to be used to input into the model for the training.

In the cyber-attack prediction step, the model predicts the status of RCV's specific operation (e.g., wheel angle, speed, etc.) based on the past data and current input data.

In the cyber-attack detection step, the model detects cyber-attacks by comparing the predicted data which is the result from the prediction step and the real input data that is received by the remote control center. The model sets a threshold value to make a decision on the violation of the normal operation. The model uses Long-Short Term Memory (LSTM) for time-series prediction. It is a widely

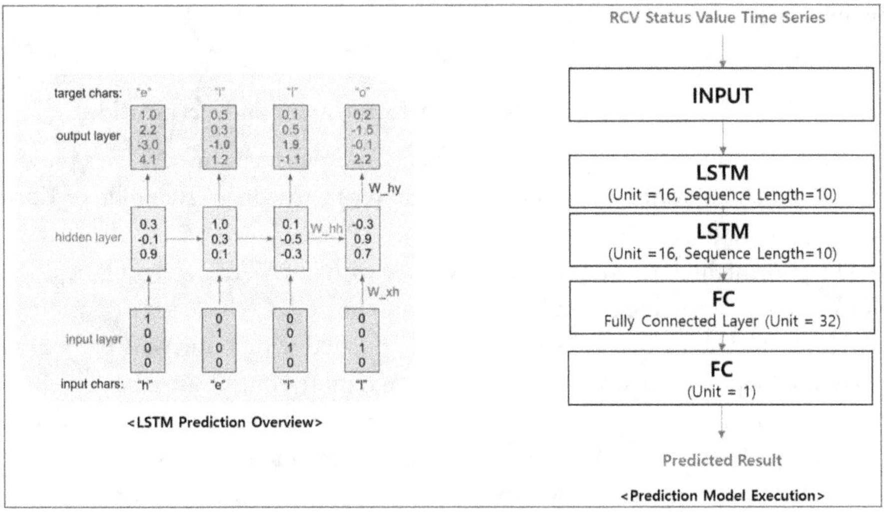

Figure 9. LSTM Execution Overview for RCV Cyber-attack Detection.

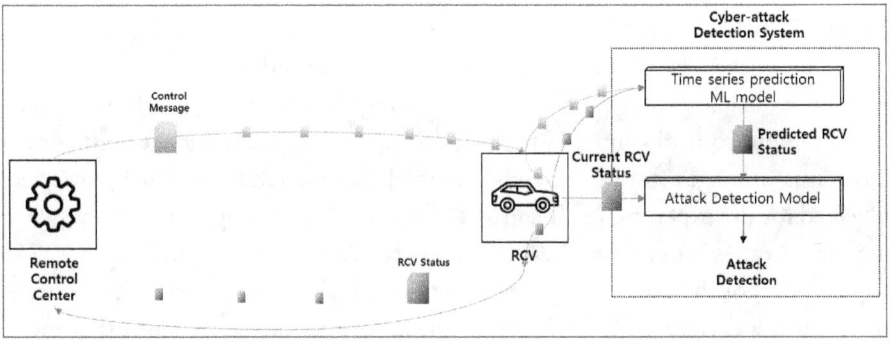

Figure 10. Overview of RCV cyber-attack detection solution.

used and proven deep learning architecture based on an artificial recurrent neural network (RNN) and applied for problems involving sequences and time series such as unsegmented, connected handwriting recognition, speech recognition, machine translation, robot control, video games, and healthcare.

Figure 9 depicts how LSTM typically performs its regression prediction based on the time-series data inputs and how LSTM model executes RCV operational status messages to predict its results based on the current and past input data.

Figure 10 illustrates our RCV cyber-attack detection system based on the functional architecture and the detection model described above. Control messages and RCV status messages are exchanged between the remote control center and RCV. They are collected and sent to a cyber attack detection system which is predicted by the LSTM model generated by the AI orchestrator in the AI server. The output of the prediction process and actual current RCV status data received are compared by the attack detection model which is a part of the inference engine. Cyber-attack is decided based on the threshold set by the system.

6. RCV Threat Detection and Prediction Experiment and Verification

We have performed RCV cyber-attack detection experiment in KATEC's proving ground which is located in KATEC headquarter in Cheonan, Republic of Korea during June of 2022.

The implemented remote controlled vehicle with mmWAVE system is deployed on a KATECH's proving ground for the remote controlled vehicle use case as shown in Figure 11. The remote control vehicle is equipped with a mmWAVE OBU on top of the loop and is used to test and validate remote-control use case. The RCV equips a total of 8 cameras (front, left, right, rear, and 4 cameras for surround-view monitoring) for real-time video streaming to the remote control station through mmWAVE communication. A remote control station was developed using real automotive parts. The mmWAVE base station and core network are mounted in the moving base station vehicle, a Renault Master van with a remote control station. During the field trial, the key functionalities (real-time video streaming and control RCV via mmWAVE communication) are tested and validated.

We installed an AI edge module in the testing RCV and also an AI orchestrator in the remote control center. While RCV is operating, we collected both control and status messages from both ends. Figure 12 shows data sets which have been collected for the experiments. It consists of two control message sets and five wheel angle execution status message sets. They are named 600 h/701 h and 720 h~724 h respectively. For the experiment, we only collected wheel angle data at this time and other status messages (e.g., RCV speed) collection and their associated experiments are scheduled in the near future.

Figure 11. Field trial of the remote controlled vehicle use case via mmWAVE communication in KATECH testing ground.

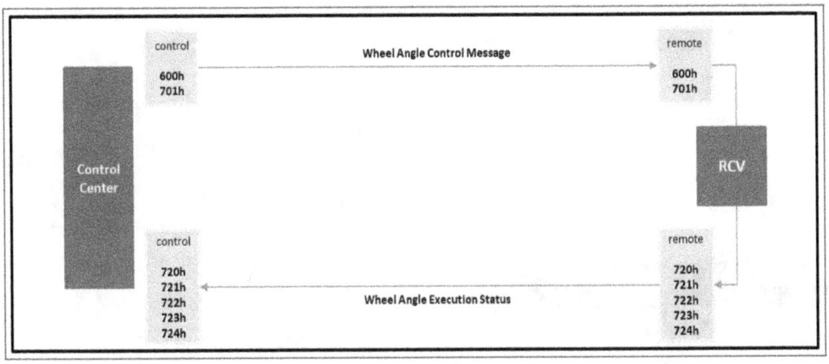

Figure 12. Data sets for cyber-attack detection experiment.

- Control Messages
  - 600h (Wheel Angle Control)
    - **APM_nSteerAngle** **(Float)**
    - APM_AutoManualSignal (Boolean)
    - FlashingIndicatorCTRL (Optional)
    - APM_nAlive_Cnt (Int)
  - 701h (Speed Control)
    - **nAcc_SetPos** **(Float)**
    - ASM_AutoManual (Boolean)
    - nBRK_CtrlState (Boolean)
    - nESTOP (Boolean)
    - **nBrake_SetPos** **(Float)**
- RCV Status
  - 720h
    - **SteeringWheelAngle** **(Float)**
    - **VehicleSpeed** **(Float)**
    - ...
  - 721h
  - ...

Figure 13. RCV control and status messages and attributes.

Figure 13 describes details on the messages and their associated attributes. The control message 600 h is to control the wheel angle of RCV. It includes various attributes for wheel angle control. In our experiment, we are mainly interested in APM_nSteerAngle which represents the angle of the wheel. The control message 701 h is to control RCV status. It also includes various attributes associated with RCV speed control. For our experiment, our main interests are nAcc_SetPos and

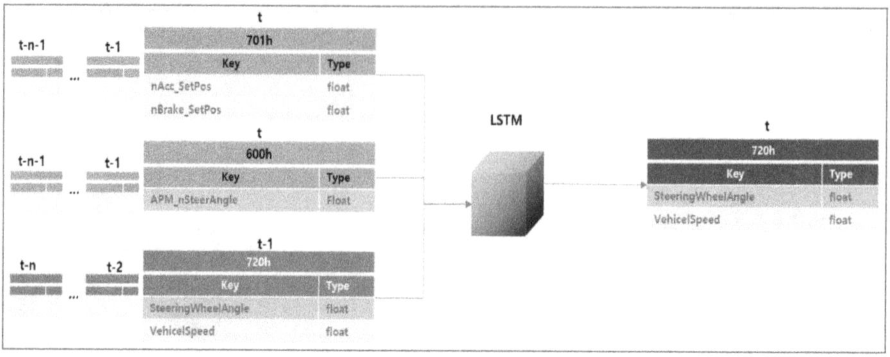

Figure 14. LSTM model training for RCV cyber-attack detection.

nBrake_SetPos. The RCV status message 720 h contains the actual operational results of both SteeringWheelAngle and VehicleSpeed.

Figure 14 describes how LSTM model training is performed to generate RCV cyber-attack detection LSTM model which can be deployed in the inference functional element in both AI edge and server. The input data consists of a series of time t and past up to n. At each time t, a vector includes five data: 600 h APM_nSteerAngle and nBrake_SetPos, 701 h APM_nSteerAngle, and 720 h SteeringWheelAngle and VehicleSpeed. In our experiment, we currently use two wheel angle data only. The input data is comprised of a time-series of the vector which has the five vectors above. The dimension of the data is three dimensions (the number of samples, 10, 5) if n is assumed to be 10. In our experiment, the actual dimension is the number of samples, 10, 2). The LSTM model performs training based on the above input data. The resulting trained model is supposed to predict RCV status of time t, that is, 720 h SteeringWheelAngle and VehicleSpeed.

Figure 15 provides the result of the experiment that we conducted. We generated three LSTM models by executing training of three data sets. Verification is performed among three models and data sets interchangeably. The row represents data sets which were used in the model training and the column represents data sets used for the verification. For example, the prediction value 0.0083 in the table (1,1) represents the result of verification of the same data set which is used to train the data of 06/14 17.28. The prediction value 0.0052 represents the result of verification using data set of 06/14 17.34 over data set 06/14 17.28 which is used to train the model. The result can be found as shown in the table. The prediction value generally converged into zero which means that the model behaves as expected with accuracy. However, there are still some gaps in the predicted value since the distribution of data sets cannot be identical. We consider these gaps as normal which can occur in the normal RCV operational situation. The criteria that we set to decide a threshold value to detect cyber-attack among them the largest value which means

- The experiments results is converging to 0
- Which means that our LSTM based prediction model can predict steering wheel angle as expected
- The largest loss data in the dataset is used as a threshold to detect a cyber-attack: In our case 0.051 is the one for this condition and will be used as a threshold value

<Prediction Experiment Result>

Measurement Unit : MSE		DATASET		
		06/14 17:28	06/14 17:34	06/14 17:38
MODEL	06/14 17:28	0.0083	0.0052	0.01
MODEL	06/14 17:34	0.027	0.0029	0.051
MODEL	06/14 17:38	0.0112	0.013	0.004

Figure 15. RCV cyber-attack prediction and detection experiment result.

that the difference between predicted value and the actual observed value is largest. In our experiment the value is 0.051 as shown in Figure 15. Therefore, we can detect RCV cyber-attack when the observed value is higher than 0.051.

Conclusions and Future Work

We described RCV cyber-attack detection requirements, functional architecture, detection mechanism based on machine learning technology, and initial experiment and verification results. The result of our research is still at an early stage and limited and further experiments are planned in the upcoming next 12 months since Pillar 4 is still on-going for the next 12 more months. Our future research focus will extend the coverage of RCV operational status besides SteeringWheelAngle such as VehcleSpeed, etc. We are going to further explore any additional ML algorithms and models to improve accuracy and performance for RCV cyber-attack detection.

Conclusions

Self-driving and connected vehicles are rapidly evolving and will eventually become the main mode of transportation for people and cargo. While from a mechanical point of view little has changed compared to traditional cars, the software is at the core of this rapid transformation. These new capabilities allow for use cases, products, and applications that were unviable a few years ago, but also surface problems that require careful research and design. Vehicles, whether autonomous or not, remain a safety-critical application that must guarantee the safety of their passengers. The increasing integration of software and the connectivity of vehicles with their environment creates attack surfaces that were previously unknown and must be carefully studied. It was CARAMEL's goal to address some of those challenges, provide advanced methods to mitigate them and explore further problems that are still to be solved.

In this work, we presented the different contributions of the CARAMEL project. In section 1 we showcased the different modules and results of pillar 1 (Autonomous Mobility) focusing on developing innovative technologies to detect attacks against the sensors in the vehicle, e.g. attacks using generative adversarial networks to disturb the object detection algorithms in the car. Subsequently, a novel intrusion detection system – the anti-hacking device – was integrated directly into the vehicle. The anti-hacking device uses machine-learning technology to detect attacks. Due to its unique design, it made it easy and safe to update these detection algorithms regularly to counter novel attacks in a short time. This research made autonomous or semi-autonomous driving more secure against advanced attack scenarios and will drive the acceptance and adoption of these innovations by the general public in Europe. In section 2 we showcased the modules and results of pillar 2 (Connected Mobility) focusing on developing advanced attack detection technologies for the connected vehicle, especially the V2X protocols between the car and other cars and roadside units. Additionally, partners implemented innovative threat detection mechanisms for direct integration into onboard units (controllers built into the car). Furthermore, the project developed an innovative bridge technology to make competing 802.11p and C-V2X networking technologies compatible using MEC devices or roadside units. In section 3 we showcased the modules and results of pillar 3 (Electromobility) where we implemented security mechanisms to protect against wide-scale attacks targeting the European EV charging infrastructure: Charging stations are part of the European critical infrastructure, and attacks that cripple this infrastructure could adversely affect the flow of passengers and goods in Europe, the project implemented machine learning models to detect attacks on the level of the whole eCharging backend infrastructure so that early mitigations can be implemented. Finally, In section 4 we showcased the modules and results of pillar 4 (Remote Control Vehicle) where the Korean partners, who still has one year of project based on Korean funding, continued working on the implementation of the remote control vehicle based on the mmWAVE (23 GHz) use case, building a data processing architecture and developing a Malicious Traffic Detection Solution LSTM (Long-Short Term Memory)-based cyberattack anomaly prediction/detection.

The future of mobility is clearly delineated by the increased support of capital markets, the high demand for private cars in emerging markets (e.g. China, India, etc.), the global expectations to hold down climate change, and the possibilities that open up by the application of new technologies, available from now to automotive industries. Cybersecurity is a defining aspect of the automotive and mobility ecosystem in the digital age, mainly by its purpose to maintain awareness of risks and threats and its operational capacity to ensure defenses against attacks on our privacy, safety, and well-being during driving and riding. The most pivotal observation of this analysis is the universal acknowledgment of Cybersecurity as a critical factor in automotive. Consumers, OEMs, and regulators all contribute from their standpoint to a heightened awareness of Cybersecurity issues, along and across the automotive domain. Although, those different actors perceive and present different levels of cybersecurity awareness they all agree that we now have an additional layer of concern, risk, and threat to deal with while being in the production and operating environment of autonomous, connected, and electromobility. Digital transformation, with its disruptive technologies in Artificial Intelligence & Machine Learning; high-performance networks & connectivity of anything; integration & efficiency of sensors; big data analytics & knowledge management; together with the prospect of electromobility being the dominant scenario of future mobility; all contribute to enlarge and multiply the attack surface that vehicles will be exposed from now on to all kinds of intentions and involuntary acts.

It is also apparent that the local-minded, one for all type approval process of to-be-sold devices and vehicles is insufficient. The UNECE 155, for example, advises a Cyber Security Management system to all stakeholders for good reason. So far, the law is not implemented. Some of the solutions presented in CARAMEL are backend oriented, so it could be considered as a "Seeing is believing" style, since the implementation of cross-stakeholder, "online" messaging about threat awareness is hampered by political issues within and between the industrial players, fragmentation of the public players, unclear business case.

This has created an awareness and momentum for cybersecurity in the automotive industry and market, where actors are willing to take on the complex challenges that come along with autonomous, connected, and electromobility in the digital age. Followed up by considerable investments in research and innovation, transformation efforts to get the right mix of people, technologies and methodologies, expressed willingness to cooperate and find synergies, and above all the readiness to meet efficiently and effectively the security, safety, reliability, and compliance requirements in the automotive industry. The appraisal of the status and prospects of the mobility market has revealed the particular conditions that describe and set the playground for industrial, commercial, and operational activities in the automotive domain in Europe. At the political level Cybersecurity is institutionally acknowledged, at the highest level (European Commission), with a new dedicated strategy, an empowered specialized agency (ENISA), ample support for research & innovation, and a multidimensional regulatory, administrative, and consultancy framework that provides reference and guidance to industrial and commercial activities with respect to social and environmental responsibilities. The economics of the EU, regarding growth conditions, size & metrics of the market, demographics & institutions, resources & sales, as well as funding & costs, all favor and support practically the integration of cybersecurity practices, products, and CARAMEL services in the automotive domain, who is expected to develop in full and by large numbers in the next decades. Socially there is a clear awareness, albeit with varying specific gravity, of the critical role that cybersecurity plays in future mobility, concerning passenger & road safety, sustainability & inclusiveness, ethics & culture, and defenses against criminal activities and damages of all kinds. Technologically, the powerful momentum of digital transformation, together with other disruptive technologies and challenges provide an operating environment for adapting and adopting industrial, commercial, and business innovations to the high-tempo of developments in the automotive domain. The legal and regulatory framework for engaging in cybersecurity activities in mobility is a dense and evolving set of EU directives, UN regulations, and industrial & business standards, related to national legislation and administrative practices, which defines the rights and obligations of all involved actors and stakeholders. Environmental awareness becomes an overarching principle in automotive since EU and global decarbonization targets together with Climate Change strategies are expected to be benefited by restraining car emissions and the road & transport impact altogether through electromobility, digital transformation of manufacturing, and efficient usage of vehicles. Cybersecurity becomes a holistic concept that runs throughout the digital and physical continuum of automotive mobility. A design and function requirement that simply cannot be ignored because of the potentially grave implications in human life & health, individual well-being & social community, economy & production, and at first most consumers & markets. The eco-system is in need, as the consortium of CARAMEL well understands, of a comprehensive cybersecurity strategy & practices, engaging directly and throughout the product, software, and services development life cycle and thereafter enabling a vigilant eye on the functions and operations of systems, platforms, and networks.

CARAMEL advocates to design, developing and operating a risk-aware, process-driven solution that integrates cybersecurity, as an end-to-end concept, that has to be taken into consideration in safety provisions & regulations, engineering & administration, manufacturing & assembly, and usage & services. Moreover, the CARAMEL consortium brings along a range of complementary multidisciplinary areas of expertise, including direct knowledge of autonomous, connected, electromobility, and cybersecurity and the explicit willingness to cooperate in a synergistic fashion under the banner of catering to the cybersecurity of the whole domain. The more the path to collaborative threat awareness is laid out the more we hope to expect actual implementation across the cloud and subsequently also for companies active in the EU ecosystem like Capgemini and Atos (two partners from CARAMEL's Consortium) who mostly perform like integrators between the sectors, partly via the cloud. CARAMEL can be Europe's springboard to the next level of cybersecurity for autonomous, connected, and electromobility by employing our collective resources in industrial capacity, technological prowess, social responsibility, and market leverage.

Index

Contributing Authors

European Partners

i2CAT – Sergi Mercader, Sergi Sánchez

CAPGEMINI: Eduardo Cervantes, Daniel Fulger

8Bells: Ioannis Giannoulakis

UBIWHERE: Rita Santiago

CYBERLENS: Nikos Argyropoulos

GREENFLUX: Bob Elders

SIDDROCO: Anastasios Lytos

0 INFINITY: Anish Khadka

UCY: Christos Kyrkou, Christos Laoudias, Theocharis Theochrarides

UPAT: Andreas Kloukiniotis, Andreas Papandreou, Aris S. Lalos, and Konstantinos Moustakas.

NEXTIUM by Idneo: Daniel Baños, Natalia Porras

AVL: Anusha Karemane

Korean Partners

MOBIGEN: Ys Lee

About the Editors

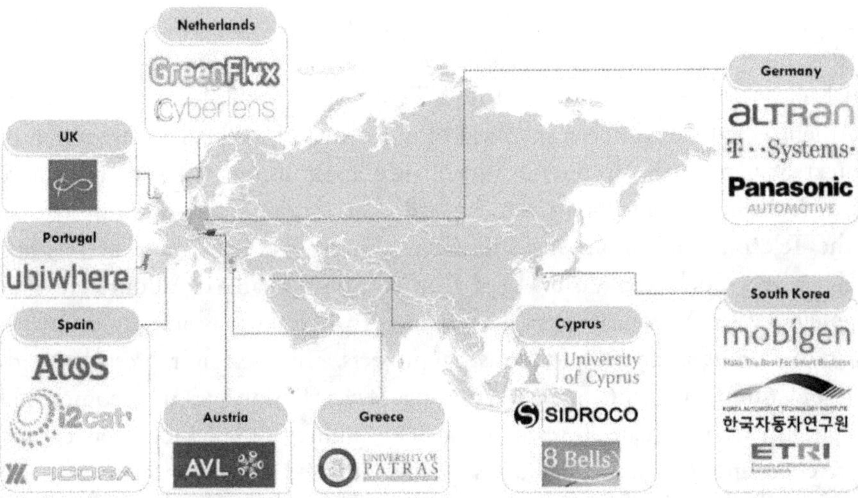

Jordi Guijarro Olivares is working as the Cybersecurity Innovation Director at i2CAT. With more than 20 years in ICT sector, he is an expert in cloud oriented and cybersecurity services. In cybersecurity arena he has experience in managing CERT/CSIRT teams employing proactive, reactive and value-added security services such as perimeter protection, infrastructure hardening, intrusion/anomaly detection, incident handling, etc. He also combines his work at i2CAT with the participation in European and national projects, and collaborates in high education programs with some universities (UPC, UOC, VIU). He obtained his bachelor in Computing Engineering at the Open University of Catalonia (UOC) and a Master in ICT Management at Universitat Ramon Llull (URL).

Peter Hofmann is senior IT security consultant at T-Systems/Telekom Security. After working for several years in the area of Mobile Security (especially Mobile Wallet) and contributing to several GSMA papers, standards, and activities, he expanded his area of expertise into the Automotive Security domain. He worked on several projects and project proposals with the three major German car manufacturers that concentrated on the mobile security/backend connectivity security aspects of the connected car concept.

Petros Kapsalas received his Diploma degree and his M.A.Sc. degree from the Technical University of Crete (Dep of Electronics & Computer Engineering) and his Ph.D. from the Electrical & Computer Engineering Department, National Technical University of Athens, Greece in 2004, 2006 and 2011, respectively. During his PhD studies he had been working as a researcher at Image Video and Multimedia Analysis Lab (ICCS, NTUA) at National Technical University of Athens where he had been involved in numerous FP7 research projects on the context of Computer Vision for Surveillance applications, visual datasets automated Indexing. From 2004–2006 he had been working as a research assistant in the Telecommunications Institute Crete. From 2011 until now he has been working as a Computer Vision Expert on two highly ranked Automotive Tier1 Suppliers (Valeo Vision Systems & Panasonic) where he was actively participating in numerous academic and industrial projects. His research interests lie on the fields of Computer Vision, Signal Processing, Sensor Fusion and 3D Geometry. He has published more than 20 research papers in numerous scientific peer-reviewed conferences and Journals while he has also been invited speaker on many Scientific and Industrial Conferences, lead committee member of the IEEE P2020 Automotive Standard and Guest Editor on Journal's Special Issues. He has participated as a researcher in numerous EU-research projects: OPTAG, IMAGINATION, MESH, EUROPEANA, X-MEDIA and he acts as a reviewer for many Computer Vision Conferences and Journals (ICIP, BMVC, ACM, etc.).

Jordi Casademont received the M.Sc. and Ph.D. degrees in telecommunications from Universitat Politècnica de Catalunya (UPC) in 1992 and 1998, respectively. He is an Associate Professor with the UPC and collaborator of i2CAT Foundation. He is an active member of the Wireless Networks Group (WNG) and works in the fields of networking and MAC mechanisms, mesh networks, wireless sensor networks and vehicular networks.

Saber Mhiri is a senior researcher at i2CAT Foundation. He has a master's degree from the higher institute of computer science and multimedia of Gabes (ISIMG) in 2014. He obtained his Ph.D. degree from the University of Vigo after preparing a thesis on Optimization Access Point Assignment Algorithms in SDN-Controlled Wireless Access Networks with the GTI team in 2019. At i2CAT Foundation, he contributes to the research activity of the cybersecurity department and is actively involved in collaborative research projects. His research interests include IoT and CCAM cybersecurity as well as network security.

Nikos Piperigkos received the Diploma and M.Sc. degrees from Computer Engineering and Informatics Department (CEID) of University of Patras, Greece in 2018 and 2020, respectively. Currently, he pursues his PhD degree at the same department. Since 2017, he is a member of Signal Processing and Communications Lab of CEID. He joined the Industrial Systems Institute, Athena Research Center and became member of Multimedia Information Processing Group in 2019. He has participated in two Horizon 2020 R&D projects. His research interests include cooperative localization and tracking, sensor fusion, intelligent transportation systems, distributed estimation, adaptive signal processing and learning algorithms.

Rodrigo Diaz is the Head of the Cybersecurity Unit inside Atos Research and Innovation, the R&D hub of Atos. He has been working in the Cybersecurity domain for more than 15 years. Rodrigo is member of the Atos Scientific Community which aims at crafting the Group's vision for the future of technology in business, and anticipating the upcoming trends and technologies that will reshape businesses and society in the years ahead. He is also member of the Atos Expert Community contributing to the Cybersecurity domain.

Bruno Cordero is a research engineer at i2Cat, in Barcelona. Currently working on the Mobile & Wireless Networks Research Unit, his main topics include Satellite Communications optimization techniques and V2X technologies. In 2018 he received his Bachelor degree in Telecommunication Networks from the Pompeu Fabra University (UPF).

Jordi Marias Born close to Barcelona, with a Bachelor's and Master's Studies in Telecom Engineering from the Universitat Politècnica de Catalunya (UPC). And the mobility period as a visiting researcher at the Illinois Insititute of Technology

(IIT) during my Master's Thesis; allowed me to start my research career by presenting my first paper at the IEEE Globecom 2022.

I'm currently working as a R&D Network engineer at I2Cat Foundation further developing the V2X (Connected Vehicle) technologies. My position also requires professional software development experience I acquired working for Roche Diagnostics for around 2 years.

Adrián Pino is a Telecommunications Engineer at i2CAT Foundation in Barcelona, Spain. He received his B.Sc. (2019) in Telecommunications Engineering (Telematics) from the University of Granada (UGR) and he received his M.Sc. in Advanced Telecommunication Technologies Engineering (5G Networks) from the Polytechnic University of Catalonia (UPC) in 2021. His current work/research interest is focused around Container Technology, NFV, Edge Computing and Cloud Native Network Orchestration in the context of 5G and B5G networks. He has experience in Cloud Integration, DevOps methodology (CI/CD) and network testbed setup.

Theocharis Saoulidis holds a Master's degree in "Web Intelligence" from the International Hellenic University (IHU). He has published research papers in important conferences such as IEEE ICC. His major research interests include Machine Learning, Software Defined Networks, Network function virtualization, Internet-of-Things and Cybersecurity.

Josep Escrig is the director of the Distributed Artificial Intelligence research area of I2CAT. He is an Industrial Engineer and MEng in Sustainability and Energy Efficiency by the University Jaume I (Spain); MEng in Mechanical Engineering by the INSA Toulouse (France) and PhD by the University of Nottingham (UK). During his PhD, he also worked as a visiting researcher in the SINTEF Petroleum Labs in Trondheim (Norway). After his PdH, Josep became the principal data engineer of the TMF Consortium (Imperial College of London). Later, he worked as a Research Fellow at the University of Nottingham conducting research and consultancy in AI for the manufacturing industry (Innova UK program). During these years, Dr Escrig published +10 papers in high impact peer reviewed journals. In 2019, Josep joined i2CAT as the DAI research area director, and since then his area has been involved in +5 H2020 research projects, and +15 projects with private companies and public institutions.

Choi You Jun/Ph.D is the Lab. leader of Mobility Safety Lab./Automated Driving Tech. Research Division/Korea Automotive Technology Institute, Ph.D in Transportation Engineering (2018) and ISO TC204 SWG17.1 Convenor.

Taesang Choi/Ph.D received his MS (1990) and Ph.D (1995) degrees in Computer Science and Telecommunications in Univ. of Missouri-Kansas City. He joined ETRI in 1996 and is currently working as a principal research staff. He has been actively involved in the R&D of Traffic Engineering, Traffic Measurement and Analysis, SDN/NFV management, 5G network slice management, and Quantum Key Distribution Network (QKDN) control and management. He has also actively contributed in various SDOs and open source activities such as IETF, ITU-T, ONF, ONOS and etc. He is currently acting as ITU-T SG13 Question 4 Rapporteur and International IT Standardization Expert representing Republic of Korea

www.ingramcontent.com/pod-product-compliance
Lightning Source LLC
LaVergne TN
LVHW010720210926
848485LV00002B/85